PARA MI QUERI
HIJA, CON UN GRAN
BESO.   RONI.

How Nations Innovate

How Nations Innovate

The Political Economy of
Technological Innovation in
Affluent Capitalist Economies

Jingjing Huo

OXFORD
UNIVERSITY PRESS

OXFORD
UNIVERSITY PRESS

Great Clarendon Street, Oxford, OX2 6DP,
United Kingdom

Oxford University Press is a department of the University of Oxford.
It furthers the University's objective of excellence in research, scholarship,
and education by publishing worldwide. Oxford is a registered trade mark of
Oxford University Press in the UK and in certain other countries

© Jingjing Huo 2015

The moral rights of the author have been asserted

First Edition published in 2015

Impression: 1

All rights reserved. No part of this publication may be reproduced, stored in
a retrieval system, or transmitted, in any form or by any means, without the
prior permission in writing of Oxford University Press, or as expressly permitted
by law, by licence or under terms agreed with the appropriate reprographics
rights organization. Enquiries concerning reproduction outside the scope of the
above should be sent to the Rights Department, Oxford University Press, at the
address above

You must not circulate this work in any other form
and you must impose this same condition on any acquirer

Published in the United States of America by Oxford University Press
198 Madison Avenue, New York, NY 10016, United States of America

British Library Cataloguing in Publication Data

Data available

Library of Congress Control Number: 2015931216

ISBN 978–0–19–873584–7

Printed and bound by
CPI Group (UK) Ltd, Croydon, CR0 4YY

Links to third party websites are provided by Oxford in good faith and
for information only. Oxford disclaims any responsibility for the materials
contained in any third party website referenced in this work.

To August

Acknowledgments

This book discusses a wide range of patterns in technological innovation in affluent capitalist economies. Chapter 2 draws from "Insider and Public Information in Varieties of Capitalism" in *Socio-Economic Review* 12(3): 489–515 (by permission of Oxford University Press); some parts of Chapter 3 draw from "The Political Economy of Technological Innovation and Employment" in *Comparative Political Studies* 43(3): 329–52 (coauthored with Hui Feng); Chapter 4 draws from "The Role of Interfirm Networks in Technological Innovation and Education" in *Comparative Political Studies* 42(5): 587–610, and some parts of Chapter 6 draw from "From Industrial Corporatism to the Social Investment State" (coauthored with John D. Stephens) in *Oxford Handbook on Transformation of the State*, ed. Stephan Leibfried, Frank Nullmeier, Evelyne Huber, Matthew Lange, Jonah Levy, and John D. Stephens (by permission of Oxford University Press).

In the course of writing this book, I have benefited disproportionately from the help and generosity of many scholars. My gratitude goes first and foremost to Professor John D. Stephens, under whom I studied as a graduate student at the University of North Carolina. It was Professor Stephens who inspired me to search for a common theme from the otherwise disparate and disjointed investigations of technological innovation I have carried out, and build them into one concrete product. In other words, Professor Stephens' genuinely unique theoretical vision gave me the crucial "eureka moment," which opened up the possibility of writing a book about technological innovation. As the book manuscript progressed, Professor Stephens unfailingly provided insightful and timely guidance whenever I needed it, reading various drafts and parts of the manuscript in great detail. Furthermore, Professor Stephens gave me the invaluable opportunity of coauthoring an article on corporatist industrial transformation in the forthcoming *Oxford Handbook on Transformation of the State*, which I draw upon in the book's concluding chapter. From start to finish, Professor Stephens' devotion to his students has been instrumental in the book's creation.

I would also like to thank Professor Peter Hall, who generously provided me with an opportunity to try out some initial empirical findings of the book, in one of the most prestigious seminar forums a comparative political economy

Acknowledgments

scholar can aspire to: the Seminar on the State and Capitalism since 1800, at the Center for European Studies, Harvard University. As the seminar's discussant, Professor Gunnar Trumbull offered extremely insightful comments on my paper about information and capitalism, which later became a crucial theme running throughout this book. Besides Professors Hall and Trumbull, I also thank the many other participants of the seminar, who provided numerous useful suggestions on how to build from the idea of technological innovation, information, and varieties of capitalism. I am also grateful to Professor Hui Feng at King's University College, University of Western Ontario, with whom I coauthored an article on employment and innovation. Professor Feng gave me the invaluable insight of understanding technological innovation through the lens of product and process innovation, which became an important theme in Chapter 3 of this book.

I would also like to thank the two anonymous reviewers for Oxford University Press. Their extremely detailed, constructive, and insightful comments, running up to fourteen single-spaced pages, opened up many avenues of improving the book which I could never have imagined on my own. The two reviewers gently guided me to connect the book to a far broader audience than I initially envisaged, including sociology, urban and regional studies, industrial economics, and case-study scholars. The two reviewers also showed me the art of engaging in a meaningful conversation with the existing literature. Their helpful advice allowed the book's quantitative analyses to become not only richer in dimension but also more nuanced in methodological choice. As a result of their help, the book was able to tell a more coherent, interesting, and relevant story about technological innovation in advanced capitalism. Many other scholars have generously offered me time, data, and suggestions in writing this book, including Professors Lane Kenworthy, Cathie Jo Martin, and Duane Swank, among others. The initial empirical results for Chapters 2 and 4 were presented at, respectively, the 2010 and 2011 Midwest Political Science Association Annual Conferences in Chicago. I thank the many panelists, discussants, and other participants who offered helpful and constructive comments on these empirical analyses.

I would also like to thank my colleagues at the Department of Political Science, University of Waterloo, who not only provided an encouraging and friendly research environment but also offered generous help and suggestions whenever I needed them. At Oxford University Press, I would like to thank Dominic Byatt, who guided the project forward with clockwork efficiency and reliability, and greeted the project with optimism and enthusiasm from start to finish. I also thank Sarah Parker for her seamless and timely communication and logistic support. In the course of writing this book, two non-technological innovations were added to my household, Frank and Douglas. I thank them for their love.

Contents

List of Figures — xi
List of Tables — xiii

1. Introduction — 1
2. Who Are Better Hunters for Innovation? — 10
3. Whose Innovation Creates More Jobs? — 59
4. Whose Innovation Creates More Inequality? — 107
5. Who Faces a Dilemma between Volatility and Output in Innovation? — 164
6. Conclusion — 213

References — 235
Index — 261

List of Figures

2.1.	IPOs and Innovation	36
2.2.	IPOs and Innovation	39
2.3a.	Venture Capital and Innovation	42
2.3b.	Venture Capital and Innovation, Before and After Onset of Crisis	48
2.4.	Bank Lending and Innovation	53
3.1.	Innovation and Employment	87
3.2.	Innovation and Employment	89
3.3.	Radical vs. Incremental Technologies	92
3.4.	Core vs. Exposed Labor Force	94
3.5.	Innovation and Productivity Level	101
3.6.	Innovation and Productivity Growth	103
4.1.	The Organization of Chapter 4	110
4.2.	Innovation and "Inequality from the Top"	150
4.3.	Innovation and "Inequality from the Top"	152
4.4.	Innovation and "Inequality from the Bottom"	153
4.5.	Innovation and "Inequality from the Bottom"	154
4.6.	Innovation and Overall Inequality	159
5.1.	Volatility vs. Innovation Output (CIS Surveys)	203
5.2.	Volatility vs. Innovation Output (CIS Surveys)	206
5.3.	Volatility vs. Innovation Output (Patents)	209

List of Tables

2.1.	OECD Countries Ranked by Coordination	25
2.2.	Poor Public Information in IPOs Finance, 1980–2002	34
2.3a.	Good Insider Information in Venture Capital Finance, 1997–2002	41
2.3b.	Good Insider Information in Venture Capital Finance, Additional Analysis	45
2.4.	Good Insider Information in Bank Finance, 1980–2002	52
3.1.	Intensity of Process Relative to Product Innovation, Twelve OECD Countries	71
3.2.	The Effect of Product and Process Innovation on Employment, Twelve OECD Countries	83
3.3.	The Effect of Innovation on Employment, Fifteen OECD Countries (1980–2006)	86
3.4.	The Effect of Innovation on Employment, Additional Analysis	91
3.5.	The Effect of Innovation on Economic Productivity	100
4.1.	Characteristics of Education Systems	126
4.2.	Percentage Change in the Odds of Overeducation, ISSP 2000–6	128
4.3.	Percentage Change in the Odds Ratio of Overeducation, Additional Analysis	130
4.4.	Percentage of Workforce Overeducated for Given Occupations (ISCO88 Two-Digit), 2000–6	134
4.5.	The Effect of Overeducation on Patents from Radical Innovation, 2000–6	140
4.6.	The Effect of Innovation on Earnings Inequality, 1985–2001	149
4.7.	The Effect of Innovation on Earnings Inequality, Additional Analysis	158
5.1.	Ratio of Intramural over Extramural R&D, Twelve OECD Countries	184
5.2.	The Effect of Volatility Reduction on Innovation Output (Community Innovation Surveys), 1998–2006	201
5.3.	The Effect of Volatility Reduction on Innovation Output (Patents), 1998–2006	208

1

Introduction

On the special occasion of its fiftieth anniversary in 2011, the Organization for Economic Cooperation and Development (OECD) looked back on technological innovation: "The big projects like exploring the origins of Universe or probing the workings of the brain capture our imagination, but science and technology are also about making daily life better" (OECD 2010, p. II). Technological innovation, as economists long pointed out, may have dramatic impacts on nations' economic growth and prosperity. For example, because new scientific ideas are often generated from creatively "recombining" existing ideas, every new discovery massively expands the number of ways in which ideas can be recombined, allowing innovation to drive *more than exponential* growth in knowledge (Weitzman 1998). New scientific knowledge, in other words, has a multiplier effect, amplifying small productivity gaps into gigantic income gaps between nations as well as individuals (Kremer 1993; Zeira 1998).

Despite its fundamental importance to economic prosperity, technological innovation has been an underdeveloped topic in comparative political economy scholarship, until the recent literature on varieties of capitalism (Hall and Soskice 2001; Ornston 2012; Taylor 2004; Lazonick 2007; Amable and Boyer 2001; Casper et al. 2005; Casper 2000). As Peter Hall and David Soskice (2001) point out, market-based capitalism tends to do better in "radical innovation," while strategically coordinated capitalism tends to excel in "incremental innovation." This is a groundbreaking argument. Drawing from earlier insights on the comparison of capitalism (Katzenstein 1985; Zysman 1994; Lazonick 1990; Kogut 1993; Albert 1993; Crouch and Streeck 1997), it brought technology innovation to the analytical forefront of comparative political economy for advanced industrialized countries. As a result, the choice between radical and incremental innovation has set the terms for the debate on innovation that followed in the literature.

Nevertheless, the choice between radical and incremental technologies is hardly the only conundrum facing affluent capitalist economies innovating in the twenty-first century. There are many other dimensions to Hall and Soskice's (2001, p. 1) provocative question: "How can national differences in the pace or character of innovation be explained?" Similarly, there are many other ways to understand what they refer to as the "different capacities for innovation" for different types of capitalism (2001, p. 21). This book goes beyond the traditional focus on radical and incremental innovation, and takes the comparison of capitalisms to an entirely new set of topics in technological innovation, which have so far received little discussion in the comparative political economy literature, such as the impact of innovation on jobs, inequality, financial markets, and firm organization. To motivate these inquiries, the book approaches technological innovation as a *process of distribution* in one of today's most valuable economic assets: *information*. This book, in other words, is about the *political economy of information distribution*.

The Political Economy of Information Distribution

Technological innovation is all about information. On one hand, innovation creates new technological knowledge useful for economic production; on the other hand, the very process of innovation is an example of searching under imperfect information, given the uncertain and constantly evolving nature of research and development (R&D) (Freeman and Soete 1997; Nelson 2005; Mokyr 2002). As a result, by focusing on patterns of innovation, the book sheds light on *how affluent capitalist economies differ in the way they distribute information*. As Frances Bacon (1597) noted in *Meditationes Sacrae*, "knowledge is power (*ipsa scientia potestas est*)." Information is a valuable asset in many ways. In the form of technological knowledge, it can be applied to create new products or increase productivity; in the form of human capital, it allows workers to get better jobs on the labor market. Not only can information be used as valuable goods or capital, it can also be used as strategic resources in dealing with partners and competitors alike. For example, when a firm possesses hidden information unobservable to outsiders, it may act opportunistically and exploit its partners or customers. On the other hand, the strategic revelation of information may also facilitate communication and expand the opportunity to trade for all.

Because information is a valuable form of economic asset and strategic resource, all players have a stake in how information is distributed. This book is an effort to understand how varieties of capitalism differ in the way they distribute information. While questions of distribution occupy a central place in comparative political economy, much of the existing discussion focuses on the distribution of income, earnings, or fiscal resources (Bradley et al. 2003; Iversen and Soskice 2009; Clarke and Stone 2008; Berry 2008). This book, by contrast, examines the distribution of an altogether different economic resource: information. There are many ways to study the distribution of information. A very common question in distributive politics is "who gets it" (Boix 2003; Acemoglu and Robinson 2005). For example, in the financial market, do firms send "public information" (revealing it indiscriminately to all investors) or "insider information" (revealing to some investors but withholding from others)? Besides "who" it reaches, the distribution of information also matters in "how" it reaches them. As the book will show, in some countries, new technological knowledge spreads through the economy while embodied in *new products*, but in other countries, knowledge spreads in the form of *cost-efficient production methods*. Furthermore, we can also examine the other end of information distribution, and ask "where it comes from." For example, what informs employers' hiring decisions on the job market? Does their assessment of a job applicant draw upon her level of skills, or her academic credentials? While questions above all examine the macro flow of information *across markets* (financial, product, or labor), on a micro level we can also study how the *power to exploit information* is distributed *across the relationship* between firms during innovation. For example, when one firm asks another to carry out R&D on its behalf, how much leeway does it grant its agent in exploiting its own private information?

This book is an attempt to draw out the diverse real-world implications of how varieties of capitalism differ in the way they distribute information. In the book, I examine the political economy of information distribution in four different domains of technological innovation, developing theories and evidence that yield interesting, and sometimes counterintuitive, insights for each domain. Each of these four domains operates with its own distinct type of information and focal actors.

Four Domains of Technological Innovation

Financial Markets

The first domain, examined in Chapter 2, is the *financial market*. Here, the focal actors are *investors* searching for innovation projects to finance, and the focal information is *strategic information* (often hidden) about the potential firms to be financed. In this setting, I frame the inquiry of information distribution through a question of "how firms speak": will firms broadcast information to all investors in the form of "public information," or whisper "insider information" to some while withholding it from others? This question of "who gets what" is arguably the most familiar topic in the distributive politics literature (Korpi and Palme 1998; Ansell 2010; Iversen and Soskice 2006; Beramendi 2012). In this domain, I show that firms in liberal market capitalism communicate "public information," revealing it indiscriminately to all investors; by contrast, firms in strategically coordinated capitalism communicate "insider information," revealing it to some while withholding it from others.

Whether firms communicate "insider" or "public" information has direct implications for the performance of financial markets in searching for innovation projects. Since the same piece of hidden strategic information cannot be revealed both selectively and indiscriminately, there is a tradeoff between public and insider information: the more hidden knowledge is released in one form, the less is available in the other. In strategically coordinated capitalism, firms send insider information, which makes less information available as public information. As a result, "insider investors" (banks and venture capitalists) do well, while "outsider investors" (individual stock market investors) do badly. In particular, stock investors suffer from "rat races": they have to run ever faster (by offering more investment), but the treadmill (innovation output) remains stationary. In liberal market capitalism, by contrast, firms send public information to all in the market, which makes less information available as insider information. This leads to the opposite outcome: "outsider investors" do well, while "insider investors" suffer. In particular, venture capitalists are hit by "rat races" and the banking market suffers from "credit rationing."

In short, while banks and venture capitalists benefit from rich information in strategically coordinated economies, this informational advantage shifts to the millions of households and individuals investing in stocks in liberal market economies. I refer to this effect of liberal market economies as the *minority protection effect*: by encouraging firms to reveal hidden information to all, liberal market capitalism protects the interests of minority shareholders at the expense of "major investors" whose own financial weight confers themselves greater influence and access to information.

Product Markets

The second domain, examined in Chapter 3, is the *product market* for innovation output. Here, rather than investors, *innovating firms* themselves take the center stage as the main players. In this domain, the focal information is *new technological knowledge* created from innovation, which may lead to either new products or new production processes. In this setting, I frame the study of information distribution through a very different question: "how do firms cash in" on their innovation output? In particular, do firms use their new technological knowledge to develop new products, or new methods of improving productivity? Instead of "who" gets it, this question places greater emphasis on "how" new technological knowledge spreads in the economy: is it embodied in new products, or new methods to improve productivity? In this domain, I again show a sharp difference between liberal market and strategically coordinated capitalism: innovators in the former focus on selling *new products*, while innovators in the later focus on developing *new production methods to raise productivity*. In other words, liberal market economies emphasize "product innovation," while strategically coordinated economies emphasize "process innovation."

This distinction has tangible real-life consequences, because "product innovation" and "process innovation" have very different implications for employment (Edquist et al. 2001; Antonucci and Pianta 2002; Vivarelli and Pianta 2000). While the former creates new job opportunities by introducing new product lines and opening up new markets, the latter saves labor by raising the productivity of the production process. As a result, technological innovation also has very different employment implications for different types of capitalism. In strategically coordinated capitalism, new technologies destroy jobs but enhance productivity; in liberal market capitalism, by contrast, new technologies increase employment but are less capable of lifting productivity in the economy. I refer to the job-expanding effect of innovation in liberal market capitalism as the *extensive-growth effect*. In other words, innovators in these countries push the economy on the *extensive* margin, expanding product lines, creating new markets, and generating new job opportunities. Correspondingly, I refer to the productivity-deepening effect of innovation in strategically coordinated capitalism as the *intensive-growth effect*: innovators in these countries push the economy on the *intensive* margin, deepening the productivity and sophistication in the making of existing products.

Labor Markets

The third domain, examined in Chapter 4, is the *labor market*. Here, *workers and employers* become the key players, and information about workers' *human capital* (academic qualifications and training) becomes the focal information that drives the market. In this setting, I frame the inquiry about information distribution through a question of "how firms listen" on the job market. When a job application is received, what kind of information do employers draw upon in assessing the applicant's human capital? Do they focus on how much skill she has accumulated, or how many academic degrees she has under her belt? In other words, instead of "where information goes to," this question addresses the other end of information distribution: where information comes from. In this labor market domain, I show that employers in liberal market capitalism focus most of their attention on job applicants' academic background, and furthermore, they draw such information in a peculiar way: although the real value of academic education is all in the knowledge it offers, Anglo-Saxon employers scrutinize an indicator that reveals little about the real content of education—the *relative ranking* of academic *credentials*. As a result, workers in such economies face intense pressure to "outrank" each other in academic credentials, well beyond the needs of their actual occupation. In other words, there is "overeducation" at the very top of the educational ladder in liberal market Anglo-Saxon economies.[1] By contrast, employers in strategically coordinated capitalism are less interested in "credential rankings." Instead, they pay more attention to the actual skills possessed by job applicants. As a result, workers face less pressure to outrank each other in academic credentials, and overeducation at the very top is less prevalent in such countries.

Overeducation affects one of today's most hotly debated issues: inequality. By encouraging overeducation, liberal market capitalism drives an academic educational gap "from above," by pulling the top well ahead of the average in given occupation. This drive towards top education, in turn, affects the nature of technological innovation: it increases the focus on radical technologies, which pull the top ahead of the median in earnings. By contrast, workers in coordinated capitalism face less pressure to advance in academic education, and innovation is less radical. While incremental innovation helps narrow the top/median earnings gap, it also widens the gap between the median and the bottom, noninnovative and labor-intensive, sectors.

In a nutshell, liberal market capitalism suffers from two types of "inequality from the top": an educational gap (academic attainment well above the average for given occupation), which in turn allows technological innovation to drive an earnings gap (top well above the median in earnings). Because both outcomes pull the top ahead of the rest, I refer to this inequality effect of liberal market capitalism as the *superstar effect*. This forms a contrast to the *long-tail effect* in strategically coordinated capitalism, where technological innovation widens the earnings gap between bottom and median earners.[2]

---

[1] As the book will elaborate later, its notion of "overeducation" refers to the *very upper ceiling* of academic attainment (master's and doctoral degrees). This is different from the analytical framework of Goldin and Katz (2008)'s well-known study, which placed greater emphasis on a segment of attainment closer to the average: successful transition from high school to college. As a result, there is no real contradiction between the book's argument of overeducation in liberal market capitalism (such as the US) and Goldin and Katz's finding that education in the US has fallen *behind* demand. In fact, these two accounts are complementary, bringing together different educational segments to paint a more complete picture of liberal market capitalism: too many people "fall short" in making a successful transition from high school to college, but those that make it tend to "overshoot," getting academic qualifications well above the typical for their occupation.

[2] It is important to emphasize that this is not a direct comparison of *earnings inequality* across varieties of capitalism. Instead, it is a comparison of *the effect of technological innovation on earnings inequality* across varieties of capitalism. In other words, the main outcome of interest is not the *level* of inequality, but the *slope* between inequality and innovation. For this reason, there is no real

Furthermore, as I will show, while the *superstar* and *long-tail effects* capture inequalities for *specific* segments of the earning scale, countries also differ in how innovation affects *overall* inequality across the economy.[3]

All three domains above paint "macro" pictures of information, using *markets* as a canvas to study the distribution of information. By contrast, the fourth domain, examined in Chapter 5, paints a "micro" picture of information distribution, studying how the *power to exploit information* is distributed *across the principal–agent relationship* between firms during technological innovation.

Principals and Agents

When a firm as the principal asks an agent to carry out innovation on its behalf, the agent may exploit its own private, "on-the-ground," information and behave opportunistically. In this environment, *the firm and its agent* (which may be another firm or the principal's own subdivision) become the focal players, and the focal information is *new information about the innovation process*, which the agent can hide and manipulate opportunistically. In this more micro setting, I frame my inquiry through a question of "how partners allocate the power to exploit information": does the agent have *discretion* in exploiting its own private information, or does the principal *control* the agent's actions? In the book, I show that firms in liberal market capitalism use ownership control to constrain the agent's use of private information; in strategically coordinated capitalism, by contrast, firms leave full discretion to agents.

How partners allocate their power to exploit information will directly affect the partnership's ability to manage one of the most important economic risks: volatility. When one firm supplies technologies to another, volatilities inherent to the innovation process create many chances for the agent to exploit the principal's lack of information, and behave opportunistically. As a result, *technical* volatilities faced by agents during innovation turn into *opportunistic* volatilities faced by principals. In liberal market capitalism, firms use ownership *control*, preferring to obtain technologies from divisions under their own ownership hierarchy. In other words, they rely on "rule by fiat" to constrain

---

contradiction between the book's argument that innovation *widens* inequality from the bottom in coordinated capitalism and the fact that the *level* of inequality from the bottom is generally low in such countries. Instead, these two accounts are complementary, bringing together different facets of inequality (level and slope) to paint a more complete picture of inequality in affluent capitalist economies.

[3] The terms "superstar effect" and "long-tail effect" are also used in other contexts in the literature. For example, Bar-Isaac et al. (2012) use these terms to describe patterns of horizontal competition on the product market.

their agents' use of private information. While control may reduce the principal's exposure to volatility, it also downgrades agents' motivation to raise output. Control under the ownership hierarchy undermines not only agents' monetary motivation (i.e. residual profits) but also their emotional motivation (i.e. reciprocity towards trustfulness) to raise output (Demsetz 1988; Williamson 1985; Falk and Kosfeld 2006; Bartling et al. 2012). The principal may restore motivation by reducing control, but as control slips away, the principal also becomes more exposed to volatility. In other words, firms in liberal market capitalism face a binding volatility–output tradeoff: between the need to lower volatility and the need to raise innovation output, one goal undermines the other. I refer to this effect of liberal market capitalism as the *demotivation effect*, whereby the use of control crowds out agents' motivation to raise output. By contrast, in strategically coordinated capitalism, firms can use the prospect of long-term relational commitment to discourage agents from short-term opportunism. In other words, they can reduce opportunistic volatility without taking away agents' *discretion* as independent firms. As a result, the goal of lowering volatility does not crowd out the goal of motivating agent output. In essence, affluent capitalist economies present a "dual world" of possibilities for relationship governance, represented by two distinct regimes: high and low trust. While the low-trust regime faces a binding *tradeoff* between the need to lower volatility and the need to raise output, these two objectives do not crowd out each other in the high-trust regime. In fact, as I will show, in very strongly coordinated capitalism, low volatility may actually enhance high innovation output, making these two goals complementary. Crowding-out, in other words, becomes crowding-in.

Finally, in the book's concluding chapter, I reflect on some broader theoretical and practical issues that are natural extensions from the book's main arguments. Theoretically, I discuss the role of the state in the institutional transformation of coordinated capitalism, which complements the book's predominantly "firm-centered" theories of coordinated capitalism. In particular, I discuss how the state has affected the ability of coordinated capitalism institutions to adapt to a new, knowledge-intensive, economy, drawing from the burgeoning literature on the "social investment state" (Morel et al. 2011). Practically, I use the book's core findings to draw out some new perspectives on public policy. I discuss the book's practical implications for three areas of public policy: (1) financial market regulation, (2) workforce productivity, and (3) social policy.

For example, based on Chapter 2's findings on insider and public information, is greater transparency always better for financial markets? Is it possible for European governments to foster the growth of both venture capital and stock markets, even though they rely on opposite types of information (respectively insider and public information)? Based on Chapter 3's findings about how innovation affects productivity, can new technologies help American workers overcome their skill deficit relative to European workers, and if so what kind of technologies are needed? Because workforce productivity also depends heavily on worker motivation, Chapter 5's findings about control and demotivation will shed some light on how to create an enthusiastic workforce. Furthermore, Chapter 4's findings on innovation and inequality may bring lessons about how to design welfare state policies that effectively cushion the economic impact of technological innovation. For example, because innovation in European economies widens the gap between median and low earners, a social security system that *narrows* this gap (such as the integration of labor market outsiders in Scandinavia) will address the impact of innovation more effectively than one that *cements* the insider/outsider gap (such as in continental Europe). These are just some of the book's interesting policy implications that I will explore in the conclusion chapter.

In short, this book studies *innovation as a window into the political economy of information distribution* in contemporary capitalism, and shows that *how nations innovate often has deep, and sometimes counterintuitive, implications for how they compare in many areas of socioeconomic performance*. I develop these diverse implications through a vast range of questions about technological innovation new to the comparative political economy literature. In doing so, the book draws on many different sources of data, ranging from enterprise surveys, population surveys to various measures of inequality, productivity, employment, education, financial markets, and volatility. Furthermore, drawing from the secondary literature, I also supplement the quantitative analyses with various brief case-study materials, so that the findings provide not only a broad picture of how nations differ, but also some concrete examples of the causal processes at work for the individual country, industry, and firm.

2

Who Are Better Hunters for Innovation?

Starting from this chapter, the following four chapters of the book will examine how different varieties of capitalism organize information differently, and how such difference matters across four different domains of technological innovation. This chapter examines the very first stage of the innovation process, focusing on an important problem that must be solved satisfactorily even before the actual R&D starts: financing for technological innovation. Just as innovators face the task of finding the right solution to their technological project during R&D, investors face a parallel task of hunting for the right project to bet their money on, before R&D is undertaken. Markets such as banking, public equity, and venture capital are all valuable sources of funding for firms intending to raise investment capital for new production capacities and technologies (Mayer and Vives 1993; O'Sullivan 2000; Verdier 2002; Culpepper 2005; Ornston 2012). All these are examples of external financing, where the "investor" is different from the "firm" to be financed, in contrast to firms self-financing from their own retained earnings. When external investors put money in a firm's innovation project, they may not know as much about the project's potential return as the actual firm carrying it out. As the party setting up the project, the firm may have valuable, and private, information about its prospect for success, and by occupying a "ringside seat" during the R&D process, the firm may gain further information through learning by doing, experimenting, or trial and error. External investors, by contrast, face a much murkier information environment: they have to determine the potential quality of a project before it goes under way, and they lack the direct "on-the-ground" information that the firm will gain by being at the center of the innovation process (Jaffee and Russell 1976; Cosci 1993). This relative informational disadvantage of external investors creates the need for *communication*: in order to attract investors and convince them that this is the right horse to bet, the firm has to release some of its private information to investors.

This chapter, therefore, frames its inquiry through a question of "how firms speak": will firms communicate "public information" (revealing it indiscriminately to all investors) or "insider information" (revealing to some investors but withholding from others)? How does the nature of such communication differ across varieties of capitalism? In this line of inquiry, the focal actors are the various "hunters" of innovation in financial markets, such as banks, stock investors, and venture capitalists, and the focal information under examination is *strategic* information (hidden or communicated) about the potential firms to be financed.

This issue of information communication is an important theme in the comparative political economy literature. As scholarship on the varieties of capitalism points out, advanced industrialized economies differ systematically in how they organize the interaction between business, labor, and the state. As a result, different types of capitalism exploit different institutional synergies (Hall and Soskice 2001; Hall and Gingerich 2009). These synergies have many dimensions (cooperation, conflict, insurance, training, and information), and these dimensions interact. For example, cooperation and insurance complement each other to encourage training (Iversen 2005; Thelen 2004): centralized wage concertation reinforces generous social security schemes for deeply skilled workers in continental Europe. Similarly, conflicts may pave way for cooperation: unions strategically initiate conflicts with businesses or the government in order to demonstrate their bargaining power, which eases the path of ex post negotiation (Hall and Thelen 2009; Lundvall 2013; Golden 1997). Among such institutional complementarities, the affinity between communication and cooperation is one of the first to capture the attention from scholars of affluent capitalist economies. Does strategically coordinated capitalism encourage better communication of information? Are problems of information opacity more severe in countries where businesses coordinate less? There is much evidence that cooperation between firms breeds deep communication. As scholars from various angles argue, the rich flow of information between firms is an important asset for strategically coordinated economies. For example, firms may face the temptation to poach skilled workers from each other (Pigou 1912; Stevens 1996). In strategically coordinated economies, this problem is mitigated by firms' collective ability to punish (i.e. withholding access to collective resources). Such a punishment regime, in turn, relies on effective monitoring. Besides such monitoring, collective failure in training may also be averted through deep communication between firms (Martin and Swank 2012). Along the same line, students of technological innovation argue that denser interfirm information flow in European and Japanese manufacturing enables firms to coinvest in technologies with longer horizons than their Anglo-Saxon counterparts (Lazonick 1990; Sako 1992; Graf 2006; Lane and Bachmann 1996). A similar conclusion may be drawn in the realm of firm–labor interaction: strong institutions of coordinated wage bargaining reduce monitoring costs for unions. As a result, unions resort less often to productivity-destroying strikes to verify firms' bargaining strength (Lundvall 2013; Golden 1997).

However, there is also evidence that information may flourish without strategically coordinated capitalism. For example, the reasonably deep, liquid, and active stock markets in Anglo-Saxon countries suggest that these liberal market economies can indeed broadly diffuse vast amounts of information to allocate assets and pool risks across millions of individual investors. In fact, patterns of how countries differ in their capacity to communicate information may be quite subtle. Even markets that process information of a similar technological content (i.e. venture capital and stock market financing for high-tech innovation) may have different outcomes. For example, while high-tech stock markets are important strengths for the US, the returns for American venture capital have lagged behind Europe for eight out of the first ten years of the twenty-first century, according to the European Private Equity and Venture Capital Association (2011, p. 18). Of course, different types of information may be at play: while startup firms provide their own venture capitalists with insider information, they have to provide the stock market with common information when they "go public" with Initial Public Offerings (IPOs). In other words, it is important to take the difference between types of information seriously, and provide a more nuanced understanding of how firms communicate to investors across different types of capitalism.

In this chapter, I suggest that different types of capitalism specialize in communicating different types of information. *Strategically coordinated capitalism communicates insider information but suppresses public information, and vice versa for liberal market economies.* The logic is as follows. Firms can reveal private information to investors either indiscriminately as public information or selectively as insider information. Selectively targeted insider information credibly signals intention for long-term cooperation, while indiscriminately released public information credibly reveals the lack of relational commitment. Since the same hidden knowledge cannot be revealed both indiscriminately and selectively, there is a tradeoff between public and insider information: the more hidden knowledge is released in one form, the less is available in the other. In strategically coordinated capitalism, long-term cooperation encourages firms to send insider information, which makes less information available as public information. Conversely in liberal market capitalism, more opportunistic relationships prompt firms to diversify risks by sending public information to all in the market, which makes less information available as insider information. As a result, liberal market and strategically coordinated economies end up specializing in communicating different types of messages, i.e. respectively public and insider information.

Empirically, I test this theory from an angle that explicitly draws out direct implications for the performance of financial markets: in strategically coordinated capitalism, investors relying on public information suffer from poor information; in liberal market capitalism, poor information hits those investors relying on insider information instead. When investors have poor information, financial markets cannot function properly. In equity financing, the effect of poor information on market performance is reflected in "rat races" (Akerlof 1970, 1976), where investors stay where they are (on the quality of innovation projects) even if they run ever faster (by raising the size of investment). In credit markets, poor information leads to "rationing" (Stiglitz and Weiss 1981), where investors refuse to lend even if there is outstanding demand among firms planning innovation. I provide evidence that poor information hits different types of financial markets in different types of capitalism. In markets relying on public information (public equity), strategically coordinated capitalism exhibits severe symptoms of poor information (rat races); in markets relying on insider information (venture capital and banking), the symptoms of poor information (rat races and credit rationing) shift to liberal market economies. Some of these findings about capitalism and finance directly counter our traditional intuition: although venture capital markets, as an important source of risky finance, are by far the most active in some of the most liberal market economies (OECD 1995; Lamoreaux and Sokoloff 2007), the performance of venture capital in stimulating innovation success is also *poorest* in precisely these Anglo-Saxon countries. Furthermore, this pattern is *completely reversed* for the risky capital commonly understood to be *complementary* to venture finance: the stock market. In other words, OECD countries may be caught in an "investment trap":[1] although venture capital and IPOs are symbiotic investments, better performance in one may imply worse performance in the other.

More broadly, this chapter's findings suggest that what type of capitalism creates better financial hunters of innovation (as this chapter's title asks) may depend on the financial market in question. In markets operating on insider information, strategically coordinated capitalism creates more effective hunters of innovation; in markets operating on public information, liberal market capitalism produces better hunters instead. "How firms speak" in financial markets, therefore, has genuine consequences for the ability of investors to perform effectively. In liberal market economies, investors relying on commonly available information (e.g. public equity) do well but those relying on insider information (e.g. banking and private equity) do badly. In other words,

---

[1] I thank an anonymous reviewer for Oxford University Press for suggesting this very apt term to describe the equity investment conundrum facing OECD countries. I will discuss the public policy implications of this investment conundrum in greater detail in the book's conclusion chapter.

by encouraging firms to send public rather than insider information, such economies shift the informational advantage from banks, wealth venture capitalists, or angel investors to the millions of individuals and households investing in the stock market. I refer to this effect of liberal market capitalism as the *minority protection effect*, which protects the interests of minority shareholders at the expense of "major investors" whose own financial weight confers themselves greater influence and access to information.

In what follows, I develop a theory of public and insider information in varieties of capitalism. To do so, I build on Michael Spence (1973)'s seminal theory of "product quality" signaling, extending its insight to the realm of "relationship" signaling. When firms communicate, public and insider information serve different purposes. The release of public information allows firms to *pool* risks *across the market*, while the communication of insider information allows firms to *smooth* risks *over time* with committed partners. A firm planning one-off opportunistic relationships cannot smooth risks effectively, and a firm planning long-term commitment cannot pool risks effectively. As a result, smoothing (via insider information) and pooling (via public information) are credible signals of, respectively, long-term reciprocal and short-term opportunistic relationships. Because insider and public information signal different types of relationships, firms in strategically coordinated economies communicate differently than their counterparts in liberal market economies. The former specialize in communicating insider information, while the latter specialize in public information.

Insider and Public Information in Varieties of Capitalism

Signaling and Relationships

Trading often takes place in the presence of some hidden information held by one side or the other. In other words, one party to the transaction often tends to have some private information about the product or the market which cannot be easily observed by the other. Sellers can, for instance, pass inferior products off as good ones, and firms may mislead banks about their creditworthiness. As a large literature has long pointed out (Akerlof 1970; Jaffee and Russell 1976; Stiglitz and Weiss 1981), the presence of such hidden information may hurt all parties to the relationship. For example, unable to separate good sellers from bad, an information-poor buyer may offer a price so low that it ends up driving out all but the worst sellers, resulting in "adverse selection." Similarly, a bank lacking adequate information to assess the quality of potential borrowers may simply refuse to lend, which results in "rationing" on the credit market. Because the presence of hidden information may restrict opportunities to trade for all, the information-rich side is often willing to reveal its own private information, by "signaling."

Michael Spence (1973) developed the seminal signaling thesis to solve the important problem of uncertain product quality, that is, how sellers credibly demonstrate to potential buyers that they offer high quality products. To do this, high-quality sellers send signals too costly for low-quality sellers to mimic. For example, job applicants obtain higher educational degrees to signal their productivity for employers, knowing that unproductive individuals tend to advance less at school. Similarly, firms use their debt-to-asset ratio to signal creditworthiness for banks, knowing that insolvent firms are unlikely to afford low debts. In short, the signal's content credibly reduces the uncertainty about the product's quality.

As the varieties of capitalism literature points out, a similar problem of uncertainty exists in the relationship between firms (Farrell 2009). Firms may choose to engage in either close long-term or arm's length one-off relationships with other firms, or investors. For those planning deep engagement, how do they credibly signal this commitment, and separate themselves from opportunists who pretend to be long-term partners? In other words, can signals credibly reveal the depth of intended relationship? It is the arrival of unforeseen contingencies that create the possibility for engaging in opportunism (and the alternative, commitment). Do opportunists and committed partners respond to such uncertainties with different strategies? If opportunistic and committed relationships demand mutually distinctive strategies of coping with future contingencies, then the very adoption of a strategy sends a credible signal about relationship intentions. A strategy of "smoothing" signals commitment, and "pooling" reveals the opposite.

When firms engage in long-term reciprocal relationships, risks from future contingencies are smoothed intertemporally (Allen and Gale 2000). For example, by refraining now from poaching skilled workers or launching a hostile takeover against those having invested in expensive human capital or met financial trouble, the committing firm expects to be reciprocally shielded from predation, when it itself recruits top technicians or makes losses at some future point. The similar is true with relational banking. The bank forsakes immediate claims to possible profits made by the borrowing firm (in contrast to equity financing), and in return its future claims to loan repayment is protected against possible losses made by the firm at the end of the loan period. By contrast, in one-off opportunistic relationships, firms pool the risks from future contingencies cross-sectionally (Harris and Raviv 1993; Harrison and Kreps 1978). This is most sharply illustrated in the stock market. Here, as a typical example of spot transaction, a relationship can be terminated the minute the investor becomes pessimistic about the firm's prospect and liquidates its stocks. However, because different investors have widely diverging beliefs about a firm's prospect, aggregation of diverse opinions protects the firm from an all-out investor flight: where there are pessimists, their impact on the firm is offset by optimists. The higher the uncertainty, the more widely investors agree to disagree, and the more effective such cross-sectional pooling of risks.

Firms reveal private information willingly: by doing so, they prevent outcomes such as adverse selection and credit rationing, and expand their own opportunities for trade. As I suggest next, firms may communicate private information in different ways, allowing them to either smooth risks over time or pool risks across the market. As a result, how firms communicate may credibly signal the type of relationship they intend to establish. Because strategically coordinated and liberal market economies support different types of relationships, they motivate firms to communicate private information in different ways as well. In the former, deep long-term coordination encourages firms to send "insider information" as a signal of relational commitment; in the latter, opportunistic turnovers in relationships encourage firms to send "public information" across the market.

Insider and Public Information

When firms reveal private information, they can either target selectively or indiscriminately. If the information is selectively released to some but withheld from others, it becomes insider information, privy only to certain players in the market. Insider information narrows the signal's audience, leaving the signaling firm with a shallow pool of potential trade partners. However, by "burning its own bridge," this costly method of signaling allows the firm to credibly demonstrate its deeper commitment to insiders than outsiders. This "selective giving" mechanism for credible commitment is extensively discussed in the gift-giving literature, where gifts of some genuine value are targeted selectively in sociopolitical exchanges to credibly differentiate committed partners from opportunists who offer only "cheap talk" (Camerer 1988; Carmichael and MacLeod 1997). By credibly demonstrating its commitment, the firm offering insider information may compensate for the shallowness of its partner pool with the depth of relationships it ends up building.

By contrast, the firm can also maximize the size of its audience by releasing private knowledge as public information, revealing to everyone in the market indiscriminately. Revealed indiscriminately, the information can no longer signal special commitment to a specific trading partner. Because public information does not serve as a ticket to deep long-term relationship, it does not help the firm smooth risks over time. However, by maximizing the size of the audience it attracts, public information enables the firm to cross-sectionally pool its risks across the market more effectively.

Opportunistic firms cannot afford signaling insider information because its narrow audience prevents them from pooling risks across a wide section of the market. By contrast, firms planning to smooth risks over time cannot afford public information because its indiscriminate targeting prevents them from credibly signaling commitment. In other words, the depth of relationship affects the nature of information revelation. When the relationship is dense and long term, firms send insider information; when the relationship is arm's length and short term, they send public information instead. A given piece of hidden knowledge, however, cannot be revealed both indiscriminately and selectively. The more private information is revealed in one form, the less is available in the other. The more strategically coordinated the economy, the more firms send insider information to signal relationship commitment, and the less information is available as public information, and vice versa. Strategic coordination, in other words, creates a tradeoff between insider and public information:

> *The more coordinated the economy, the better (worse) the quality of insider (public) information.*

In the next section, I draw out and test the implications of this theory in a setting where the distinction between public and insider information is especially transparent, and where the finding speaks directly to the performance of financial markets across varieties of capitalism: external financing for technological innovation.

Rat Races and Rationing in Varieties of Capitalism

As noted earlier, this chapter's core theoretical proposition is that strategically coordinated economies communicate more insider information and less public information than liberal market economies. The case of external R&D financing provides an especially fitting empirical environment to test concrete implications from this theory, for several reasons. First, unlike self-finance from retained earnings, external finance creates a *relationship*, between two *separate* players (the investor and the vested firm), which is a necessary condition for the possession and communication of private information, i.e. the core activities driving this theory. Second, even external investors may not face information opacity about the financed projects if these are well-understood and repeated tasks (such as routine production) with little uncertainty. In new tasks with greater uncertainty such as technological innovation, by contrast, the presence of private information unobservable to the investor becomes a more relevant concern.

Third, external financing is a setting where insider information and public information can be distinguished especially transparently, because different types of financial markets rely on different types of information. Although in reality no markets use exclusively only one type of information, some information is more vital for some markets than others. In both lending and private equity (such as venture capital) markets, investors' expected returns have a relatively long time frame (Hellman et al. 2008). Because of the depth of financial commitment, investors require detailed screening of potential projects and sustained monitoring of ongoing projects, which in turn relies on substantial insider information from the financed firms (Kaplan and Strömberg 2001). Therefore, lending and private equity are fertile testing grounds for the quality of insider information. By contrast, the public equity (IPOs) market exploits the heterogeneity in beliefs across millions of individual investors to even out risks and allocate assets. As a result, its proper functioning relies heavily on the flow of public information (Fama 1970; Harris and Raviv 1993). For example, Bhattacharya and Daouk (2002) find that eighty-seven out of 103 countries have stock market insider trading laws which prohibit the use of information that is not publicly available, and that enforcement of such laws considerably raises the returns of public equity. Therefore, public equity markets provide a fitting context for testing the quality of public information. Furthermore, testing this theory of public and insider information through R&D financing also has the additional benefit of addressing a substantively important topic: the performance of financial markets across varieties of capitalism.

To construct my hypotheses, I start with public information in the context of public equity finance. What happens to the public equity market if the quality of public information is poor? Among R&D projects "going public" (i.e. issuing IPOs on the stock market) to raise money, there are both good projects with high output and bad projects with low output. This outcome, however, cannot be observed at the point of investment. Of course, investors prefer good to bad projects, but they have difficulty telling one from the other, because all projects will try to pass themselves off as good ones when they peddle themselves to potential investors. Unable to separate good from bad projects, the IPOs investor is only willing to offer an amount up to the value of the *average* quality among all projects in the pool. However, by offering less than the value of good projects and more than the value of bad ones, this average bid will end up driving out the former and attracting only the latter, creating a "market for lemons" (Akerlof 1970). Investors, as a result, suffer from "adverse selection," disproportionately selecting the worst quality projects from the pool for investment. The more limited is publicly available information to separate good from bad projects, the more the investor relies on "average estimates" in setting her IPOs offer, and the more low-output projects this offer will attract, due to adverse selection. In other words, as public information becomes poorer, the contribution of IPOs to innovation output will become weaker. Since coordinated capitalism suppresses public information, it should aggravate adverse selection, and weaken the contribution of IPOs to innovation output.

> H1 (Poor Public Information): *the more coordinated the economy, the weaker the effect of IPOs purchase on innovation output.*

Hypothesis H1's prediction has direct implications for the performance of financial markets. Because available public information is limited by the use of insider information in strategically coordinated capitalism, public equity markets in such countries do not function effectively in searching for good innovations. Whatever investors offer, they get less than what they would otherwise deserve had the quality of public information been better. To offset this "adverse selection," they have to increase their offers. As public information becomes poorer, adverse selection intensifies, and the amount of money investors have to put in to offset such adverse effect on project quality also climbs. In other words, investors have to keep running (on the amount of offer) just to stand still (on the quality of innovation projects). Akerlof (1976) colorfully referred to this phenomenon of "running to stand still" as the "rat race": as the hamster wheel spins faster, the rat has to run faster in order to catch a given cheese. The rat has to expend greater efforts, not because more cheese will be produced, but because the wheel has sped up. The search friction caused by the dearth of public information creates a hamster's wheel for public equity investors: the poorer information, the more search friction, and the harder investors have to work to overcome this hurdle. As a result, investments in rat races are wasteful in the sense of "racing against the wind." In contrast to this, liberal market economies encourage the revelation of public information, and as a result public equity markets suffer less from rat races.

Equity markets are not the only arena for such wasteful rat races. Later in the book (Chapter 4), I devote considerable attention to another example of rat races, in the labor market, where workers are willing to invest in ever-higher academic degrees even though their productivity on the job remains stationary. In that context, liberal market and strategically coordinated economies again differ notably in the severity of rat races. For now, I return to this chapter's main theme, and explore the other side of the informational trade-off: insider information.

In contrast to public equity, both credit markets and private equity investment rely heavily on insider information. Since coordinated capitalism encourages the communication of insider information, it should alleviate the information opacity faced by banks and venture capitalists in financing innovation. As a result, while coordination exacerbates adverse selection for public equity, it should have the opposite effect for private equity by mitigating adverse selection: the marginal contribution of venture capital to innovation output should be *stronger* in more coordinated economies. Venture capital, of course, is not the only form of private equity. However, given the lack of systematic data for other types of private equity (such as angel or seed funding), I focus on venture capital in constructing the hypothesis.

> *H2 (Good Insider Information): the more coordinated the economy, the stronger the effect of venture capital investment on innovation output.*

The prediction that venture capital is more effective in more coordinated capitalism is surprising in light of the fact that venture capital investment by size is significantly more important in some of the *least* coordinated economies. For example, as percentage of Gross Domestic Product (GDP) the amount of venture capital investment in the UK averaged over time is more than four times the amount in continental European countries (see the following empirical analysis section for the source of data). These two accounts may be reconciled as a reflection of two different dimensions to the role of venture capital. On one hand, like all other financial instruments, venture capital searches for good innovation projects. The chances of successful search depend on the quality of insider information available. On the other hand, as a form of equity investment, venture capital gives investors a share of profits *in the event* of successful search. This, in other words, is what Oliver Williamson (1985) refers to as "high-powered incentives," which offer larger rewards for risk-taking than low-powered "fixed wage" incentives (such as interest rates charged in bank lending). Because of various institutional complementarities (such as flexible labor markets, fluid firm ownership, and portable human capital in advanced academic education), liberal market Anglo-Saxon economies engage in more radical high-risk innovations (Hall and Soskice 2001), even though their actual performance in these sectors is a subject of debate (Taylor 2004). As a result, these economies will attract large venture capital investment simply because of the higher-powered risk premium it offers. But even in these environments with very high-powered incentives, the search for good innovation should still fail if the quality of information is poor. In other words, these countries pose a "hard test" for the theory: if poor insider information hinders the effectiveness of venture capital where it is *most* attractive as a source of high-powered risk premium, the impact of insider information should only be *stronger* in other countries where high-powered considerations are less important. For this reason, in the following empirical section I also examine if the "insider information effect" can be detected even in such venture capital outliers.

While equity markets suffer from rat races under poor information, the implication of poor information for credit markets is somewhat different. Rather than suffering from adverse selection, lenders withhold credit altogether when information is opaque, resulting in "credit rationing" (Jaffee and Russell 1976; de Meza and Webb 1990; Cosci 1993). The reason why lenders are less tolerant about poor information is that, unlike equity investors, they do not share borrowers' residual profits. This creates a direct conflict between lenders' aversion and debtors' preference for risky projects. After receiving loans, debtors want to elevate the riskiness of projects ex post, since riskier projects tend to have larger returns. In the event of success, the bank is excluded from the increased profits; in the event of failure and bankruptcy, limited liabilities terms may prevent the bank from getting all the loans back. Any attempt of lenders to compensate for such risks by raising interest rates will only *worsen* the conflict of interest, forcing borrowers to engage in even riskier endeavors to offset the cost of higher interest rates.

When banks ration against innovation projects and divert more credits to routine production where information is less opaque, the marginal contribution from overall bank credits to R&D expenditure declines. Since coordinated capitalism encourages the communication of insider information, it should mitigate the extent of credit rationing against innovation, and strengthen the effect of bank credits on firms' R&D investment.

H3 (Good Insider Information): the more coordinated the economy, the stronger the effect of total bank credits on R&D expenditure.

All three hypotheses in this chapter predict interaction effects: the impact of financial markets on R&D investment and performance is contingent on the degree of coordination. For (public equity) markets relying on public information (H1), the more coordinated the economy, the weaker the power of investment in lifting innovation output. This implies that the interaction between coordination and investment should be negatively signed. By contrast, in (credit and venture capital) markets relying on insider information (H2 and H3), the more coordinated the economy, the more effective is external financing in boosting R&D investment and output. In such markets, therefore, the interaction between coordination and financing should be positively signed.

Empirical Analysis

Measuring Coordination

Comparative political economy scholarship has examined the dense cooperation of economic interests in contemporary capitalism from many different angles. The neocorporatist literature, for example, places strong emphasis on the *policy-making* process, and in particular, the wage bargaining process (Cameron 1984; Iversen et al. 2000; Mares 2004, 2006; Wallerstein 2008; Baccaro and Simoni 2010). Although the book does bring up various issues relevant to wage policies, it devotes greater attention to cooperation in the *production* process, adopting a "firm-centered" perspective on economic cooperation. This situates it more closely to the literature on the coordination of businesses and varieties of capitalism (Hall and Soskice 2001; Culpepper 2011; Martin and Swank 2012; Gourevitch and Shinn 2005; Ornston 2012), which devotes greater attention to firms (and investors) as the focal actors in economic cooperation. Such cooperation in production and investment is often referred to as "coordination" in this literature, and the strength of institutional foundations for such cooperation is what separates "strategically coordinated" capitalism from "liberal market" capitalism in this literature. As scholars in varieties of capitalism point out, strategically coordinated capitalism is based on institutional complementarities. In other words, a nexus of institutions, all connected through the firm as the focal player, work together in facilitating cooperation between firms (Hall and Gingerich 2009; Aoki et al. 1990). To properly measure coordination, it is important to directly capture the key elements of this institutional nexus. In this nexus, different institutions facilitate interbusiness coordination through different mechanisms. Some help businesses cooperate by connecting firms to each other, others do so by deepening the ties between investors and firms, and yet others use the political resources of the state to facilitate the cooperation between businesses. Next, I outline each of these three institutional mechanisms.

First, institutions may breed cooperation between firms by bringing firms together in common organizational membership. The literatures on business coordination and varieties of capitalism have identified a diverse range of institutions that may serve this purpose. These institutions are sometimes rather formal in membership, such as the Chambers of Business or *Fraunhofer* institutes in German-speaking countries (Braunthal 1965; Martinelli 1991; Crouch 1993; Graf 2006), or informal, such as small firm networks in Italy's center-north-east known commonly as the Third Italy (e.g. Emilia-Romagna, Friuli, and Tuscany) (Best 2001; Lazonick 2005). They also vary in how central authority is exercised. In some cases, control and coordination is along functional lines, as in national or regional industrial and trade associations evolving on the historical basis of guilds in continental Europe (Coleman and Grant 1988; Crouch 1993; Herrigel 1996; Streeck 2005; Lane and Bachmann 1996). In other cases, influence is vertical, based on customer–supplier ties that connect organization members (Smitka 1991; Sako 1992; Tylecote 1996; Miwa and Ramseyer 2006). These various cross-firm coordination bodies serve to reduce two important hurdles to interfirm cooperation: (1) collective action failure, where the stake of the individual firm is small and the cost of shirking can be externalized to the rest of the economy. With centralized authority and coordination capacity, these institutions aggregate the interests of multiple individual firms, increasing the stake in cooperation success and internalizing the externalities of cooperation failure (Olson 1965; Tuck 2008); (2) short-termism, where firms are not willing to cooperate because the cost of investing in cooperation is immediate but the benefit can only be realized later. Even firms with a relatively long horizon (for example, sustained by patient capital on the investor's side) may not be able to extend its horizon beyond its own lifespan. When firms in the current generation cannot fully expropriate the benefits of their cooperation (such as human capital or technological investment) which may spill substantially over to future generations of firms, the current generation's incentives for cooperation may decline. By contrast, a central institution (such as a trade federation) represents firms across successive generations, and as a result replaces the finite horizon of individual firms with an infinite horizon spanning across generations (Greif 2006). Since the trade federation fully internalizes the future benefits of cooperation, it has stronger incentives to enforce such behavior than unorganized individual firms.

Second, coordination institutions may also lengthen the horizon of firms (and hence facilitate their mutual cooperation) by deepening the commitment of their investors, through the concentration of investorship. With centralized investorship, a small number of investors own large shares, and through their predominant ownership stake, exercise decisive influence over the firm's strategies (Hicks and Kenworthy 1998; Gourevitch and Shinn 2005; Culpepper 2011). These dominant investors can be nonfinancial firms (such as production firms connected via cross-shareholding in continental European countries) or financial firms (such as banks in the Japanese "main bank" system) (Dietl 1998; Aoki et al. 2007). The centralized nature of investor control sets such economies apart from Anglo-Saxon countries, where share ownership tends to be more dispersed into small patches across many investors (Pagano and Volpin 2005). Even if these investors are large (e.g. institutional stockholders such as pension funds), their stake in the company is small. As a result, these minority investors face Mancur Olson's (2000) classic "roving bandit problem," where their stake in the productive enterprise is too low for them to refrain from opportunistic exit, abandoning the firm at the first sign of trouble. By contrast, dominant stakeholders are "stationary bandits." Since there are fewer of them each sharing a bigger chunk of the firm's actions and outcomes, lapse of monitoring efforts by any will have more pivotal consequences for the firm's performance, and for their own returns from investment. As a result, dominant investors' stake and influence in the firm is large enough for them to choose voice (i.e. intense monitoring and deep intervention) over exit.

These "firm-based" institutions noted above play a crucial role in facilitating long-term cooperation between firms, but the state as a third-party institution may also lend a helping hand, using its own political resources to further strengthen firms' coordination capacity (Busch 1999; Levy 2006; Maxfield and Schneider 1997). The state may facilitate business coordination either through government ministries with portfolio connections, or through large government procurements as financial incentives (Noble 1998). For example, as Fransman (1990) documented, the Japanese Ministry of International Trade and Industry (MITI) and state-owned Nippon Telegraph and Telephone (NTT) worked actively to lengthen firms' horizons, using policies and procurement contracts to favor very-long-term investment in leading-edge technological innovations. More broadly, following Ständestaat traditions, the state can politically underwrite interest groups' coordination authority. In such settings, state endorsement often enjoys broad and enduring political support, which stems from the consensual proportional representation electoral systems commonly found in strategically coordinated capitalism (Cusack et al. 2010; Martin and Swank 2008, 2011).

In other words, the institutional synergies that underpin strategically coordinated capitalism encompass several core elements: *cross-firm organization*, *investor concentration*, and *coordination with the state*. In order to properly measure coordination as the central concept in this book, it is important to directly capture each of these core institutional elements. For this purpose, the Hicks-Kenworthy (1998) coordination index is especially suitable, because it was designed to explicitly tap into these very three institutional dimensions of cooperation, across a large number of OECD countries. This sets the Hicks–Kenworthy index apart from the many other cooperation indices in the literature (as reviewed by Kenworthy (2003)), which focus on union organization or wage bargaining, and as a result do not adequately capture interbusiness cooperation as the core of this book's analysis. The Hicks–Kenworthy index extracts three main types of information.

First, the index gauges the extent of central business confederation authority over member firms. It gives a country a score of one, if its central business confederation has substantial authority over members and is weakly contested by competing federations, 0.5 if the central confederation exercises modest authority and/or is moderately contested by competitors, and zero if there is significant fragmentation among business federations and/or the central federation exercises little authority over members. Second, the index explores the nature of investor influence inside firms. A country receives a score of one if its corporate finance and governance mechanism is dominated by large investors who hold significant ownership shares for long periods. For example, as Hicks and Kenworthy (1998) point out, countries where most firms rely heavily on long-term debt rather than equity or are privately owned will fall under this category. If ownership is relatively decentralized but exit (i.e. investor turnover) is only modest, a score of 0.5 is recorded. Finally, if ownership is very decentralized and at the same time investorship suffers from a high turnover rate (so that investors frequently resort to exit rather than voice in corporate governance), the country receives a score of zero. Third, the index also measures the strength of state–business coordination. A country receives a score of one if there is relatively cooperative interaction between cohesive government agencies and centralized business/labor organizations. If the strength of such state–business coordination is moderate, the country is scored 0.5. If instead both state agencies and business interest group organizations are fragmented, and state–business relationship is combative and conflictual, the country receives a score of zero. The index is a sum of scores from the three dimensions and normalized to a continuum from zero to one. As will be shown later, this relatively fine-grained continuous measure provides the flexibility to accommodate differing approaches to classifying the varieties of capitalism.

To provide an overall picture of how affluent capitalist economies differ in the strength of coordination, the left half of Table 2.1 ranks eighteen OECD countries by their coordination scores on the Hicks–Kenworthy index. As the table shows, the ranking is consistent with the common understanding that capitalism is more strategically coordinated in Europe (and Japan) than Anglo-Saxon countries: the average strength of coordination among the former

Table 2.1. OECD Countries Ranked by Coordination

Hicks–Kenworthy		Hall–Gingerich	
Sweden	0.97	Austria	1.0
Norway	0.96	Germany	0.95
Austria	0.96	Italy	0.87
Finland	0.88	Norway	0.76
Germany	0.80	Japan	0.74
Japan	0.77	Belgium	0.74
Denmark	0.72	Finland	0.72
Belgium	0.67	Denmark	0.70
Netherlands	0.58	Sweden	0.69
Switzerland	0.55	France	0.69
Italy	0.44	Netherlands	0.66
France	0.40	Switzerland	0.51
Australia	0.17	Australia	0.36
New Zealand	0.14	Ireland	0.29
United Kingdom	0.10	New Zealand	0.21
Ireland	0.07	Canada	0.13
Canada	0.04	United Kingdom	0.07
United States	0.02	United States	0.00

Source of data: Hicks and Kenworthy (1998); Hall and Gingerich (2004, 2009)

group (0.72 on the Hicks–Kenworthy index) is eight times the latter group (0.09). Furthermore, the index is also rich enough to capture various more nuanced differences across countries (Amable 2003; Bonoli 1997), consistent with well-known arguments from the literature.

For example, within the Anglo-Saxon world, there is a clear difference in coordination strength between Australia and the US. This ties in nicely with scholarship on the Antipodean "wage earners' welfare state" (Castles 1985; Watts 1997; Kelsey 1993), which argues that, in Australia and New Zealand, although businesses are organizationally fragmented (similar to the US), their capacity for cooperation is indirectly reinforced by a relatively centralized tripartite wage arbitration system. Highly actively until the late 1990s, these tripartite institutions brought together the government, businesses, and labor, and delivered stable wages as a substitute for an otherwise means-tested and small welfare state. As a result, tripartite concertation sets Antipodean capitalism apart from the rest of the Anglo-Saxon world. The other rather crowded (twelve countries) world of European economies (plus Japan) also sorts itself into three distinct subgroups. First, there is a leading league of small open economies (Sweden, Austria, Norway, and Finland) with the strongest coordination (all registering above 0.8 on the index). The second subgroup consists of countries less strong but still above average in coordination. Since the mean of the Hicks–Kenworthy coordination index is 0.51, this subgroup encompasses all remaining countries outside Southern Europe. In these countries, as the related literature points out, although there is cooperation among interest groups, its scope is comparatively more limited, due either to organizational weakness (such as in Denmark or Switzerland (Huber and Stephens 2001; Esping-Andersen 1985; Kriesi 1982)), segmentation between large and small firms (such as in Japan (Whittaker 1997)), or less cooperative relationships between the state and interest groups (such as Germany, the Netherlands, or Belgium (Lehmbruch 1984; Visser and Hemerijck 1997; Jones 2008)). Third, the Southern European countries of Italy and France constitute a subgroup of their own, ranking below average in coordination strength and standing as the borderline between liberal market Anglo-Saxon capitalism and coordinated European capitalism. This again coheres well with evidence from the existing literature, which argues that these countries exhibit hybrid characteristics of both strong and weak coordination. These countries have dense informal networks consisting primarily of small firms, but the potential for interest group coordination is directly undermined by repeated state intervention (Levy 1999; Amable 2003; Culpepper 2003).

On one hand, the index extracts sufficient information to reflect these subtler differences across capitalisms; on the other hand, it makes these subtypes directly comparable to each other, along a common metric based on three institutional dimensions (*cross-firm organization, investor concentration,* and *coordination with the state*), so that the subtypes' differences can be interpreted more richly than just mutually exclusive categories. Besides the Hicks–Kenworthy index, I also use another measure of coordination in order to reinforce the robustness of my empirical findings. Although the literature has produced a large number of neocorporatism indicators, few outside the Hicks–Kenworthy measurement framework are suitable candidates for capturing "coordination" as understood in this book; that is, cooperation with firms and investors as focal players. An important exception is the coordination index calculated by Peter Hall and Daniel Gingerich (2004, 2009). This measure is a good robustness check against the Hicks–Kenworthy index for several reasons. First and foremost, its conceptual approach to "coordination" is appropriately close to this book's focus on firms as focal players. The Hall–Gingerich index taps into six institutional elements that may be organized on two dimensions: corporate governance and labor relations. The corporate governance dimension encompasses legal protection for minority shareholders, dispersion of firm ownership control, and the size of stock markets. The labor relations dimension encompasses the level (firm, intermediate, or national) of wage coordination, the strength (weak or strong) of wage coordination, and turnover on the labor market. In other words, while the Hall–Gingerich indicator pays attention to the labor side of coordinated capitalism (which is largely left out of the Hicks–Kenworthy index), it also retains sufficient focus on the firm, in its comprehensive measure of investor–firm ties and corporate governance. Second, while sharing with Hicks–Kenworthy a relatively similar conceptual approach to coordination, the computational approach of Hall–Gingerich is very different. Instead of directly using country scores on these six criteria to measure coordination strength, Hall and Gingerich first conducted confirmatory factor analysis of these six indicators (which yielded those two dimensions noted above: corporate and labor), and then normalized the factor loadings to a composite scale from zero to one as the measure of coordination strength. Third, similar to Hicks–Kenworthy, this index has also established a track record in the varieties of capitalism literature, serving as the key measure of coordination in Hall and Gingerich's well-known analysis (2009) of institutional complementarities in liberal market versus coordinated economies.

To illustrate how the varieties of capitalism are organized in the Hall–Gingerich perspective, the right half of Table 2.1 ranks the eighteen OECD countries by their coordination scores on the Hall–Gingerich index. Comparing the Hicks–Kenworthy and Hall–Gingerich rankings, there are both similarities and differences in how the two indices organize affluent capitalist economies. On one hand, the two measures agree on the broad contrast between liberal market Anglo-Saxon economies (all of which score below median on both indices) and strategically coordinated economies in Europe and Japan (all of which score above median on both indices). Furthermore, on the very upper and lower extreme of the ranking, the two indices also reach similar conclusions about who are some of the most coordinated economies (such as Austria and Norway) and some of the least coordinated (such as the US, Canada, and the UK). On the other hand, within the broad category of Anglo-Saxon and European economies, the two indices have various differences in how they organize individual countries. For example, while the Hicks–Kenworthy index ranks most Nordic countries above continental Europe in coordination strength, the Hall–Gingerich measure puts Germany well ahead of Sweden and Finland. In the same vein, while the two indices agree on the mid-level position of one Southern European country (France), they differ on the other (Italy), which is close to France on the Hicks–Kenworthy measure but close to the top on Hall–Gingerich instead. Furthermore, while Belgium registers stronger coordination than all but one Nordic countries on Hall–Gingerich, it is less coordinated than any Nordic country on Hicks–Kenworthy. Within the Anglo-Saxon world, the two indices also differ over the assignment of Ireland, which scores notably higher on Hall–Gingerich than Hicks–Kenworthy. This difference for Ireland is not very surprising, given the wage coordination under the Irish "social partnership" arrangements (Thomas 2011; Hassel 2006). This wage aspect of coordination would be captured by the labor dimension of Hall–Gingerich but left outside Hicks–Kenworthy, which does not measure the labor side of coordination. Overall, while the Hall–Gingerich index shares broad similarities with Hicks–Kenworthy in organizing countries around a firm-centered understanding of coordination, in the actual scoring of countries it also leaves sufficient distance from Hicks–Kenworthy to make itself an appropriately distinctive alternative measure.

Measuring Financial Markets

All three hypotheses in this chapter posit interaction effects: coordination alters the impact of financial markets on R&D investment and performance. Having explained in detail the measurement of coordination in the previous section, now I turn to the measurement of financial markets. The three hypotheses examine three different markets: IPOs, venture capital, and bank lending. Unless otherwise noted, all data on financial markets come from Beck et al. (2000)'s Database on Financial Development and Structure. Firms may raise external finance by offering primary issues of shares (IPOs). These "primary" issues on the stock market are different from "secondary" issues, which pass revenue from traders to traders, rather than channeling investment funding to the primary issuing firm. To construct a measure of overall IPOs purchase by public equity investors, I divide the size of the IPOs market (all

IPOs issued as percentage of GDP) by the stock market turnover ratio. The turnover ratio is a measure of the inefficiency of equity markets, defined as the ratio of total equity market size over the total value of trade in the market. The higher this ratio, the less efficient the market in facilitating trades, and the smaller the portion of market size that is turned into active trade. By dividing IPOs market size with this ratio, I derive the actual total value of IPOs purchase, as percentage of GDP. However, money raised this way does not necessarily go to actual investment: managers may also raise public equity to signal performance (O'Sullivan 2000), if shareholders are active in disciplining managers. I will later control for this confounding factor with a measure of shareholder legal protection. For venture capital, I use data on aggregate venture capital investment from Eurostat's Science and Technology Database, the yearbooks of National Venture Capital Association (US) and the Australian Private Equity and Venture Capital Association, the annual reports of Canada Venture Capital Association, and the Asian Venture Capital Journal Group. Venture capital data are in millions of national currency, and I convert them to percentages of GDP. For bank lending, I use the size of total credits (as percentage of GDP) issued by all deposit money banks to the enterprise sector.

Dependent Variables

For hypotheses H1 and H2, the dependent variable is innovation output, measured as the number of patents granted by the United States Patent and Trademark Office (USPTO) per million inhabitants. Although patent data enjoys the clear advantage of a wide "empirical scope" (consistently available across various sectors, many countries, and multiple years), it also suffers from a somewhat narrow "conceptual scope": not all innovation output is necessarily patented, especially if innovators do not face substantial risks of unintentional knowledge disclosure and spillover. This may be the case for either new techniques developed from "process innovation" in the privacy of the shop floor, or new products from "product innovation," which often relies on implicit "learning-by-using" (Malerba 1992). Furthermore, not all patents are necessarily filed for the purpose of real innovation. Sometimes, firms may have other strategic incentives for patenting, such as leveraging proprietary knowledge to enhance bargaining power (Reitzig 2004), or preempting new entrants by crowding the product space with "sleeping patents" (that is, patents that are never used or licensed) (Gilbert and Newbery 1982). In other words, patents may understate a nation's true extent of innovation. On the other hand, this limitation needs to be balanced with the fact that, in order to fully exploit the rich time-series cross-country data on financial markets, we need to measure innovation output on an analogously encompassing scope, and this is an advantage conferred uniquely by patent data.

Furthermore, to the extent that radical innovation can be more easily patented than incremental innovation, the underestimation of innovation output in coordinated economies (which specialize in incremental innovation) will bias the finding against the book's argument that venture capital is more effective in boosting innovation output in such economies.

For hypothesis H3, the dependent variable is private enterprise R&D expenditure, measured as percentage of GDP. As the varieties of capitalism literature points out, different countries may face different worlds of innovation: while liberal market economies concentrate on high-tech radical technologies, strategically coordinated capitalism may focus more on incremental innovation in more mature technologies, although there are ongoing debates about the conclusiveness of empirical evidence for this pattern (Hall and Soskice 2001; Ornston 2012; Taylor 2004; Akkermans et al. 2009). As a result, some readers may argue that financial markets in different countries should be evaluated in the informational environment most relevant to these countries' world of innovation. To distinguish between "cutting-edge" and "mature" technologies, I use the OECD's classification of technologies (Hatzichronoglou 1997). Using three criteria (production, value-added, and technologies embodied in the R&D process), the OECD Secretariat classifies the extent of innovation into the following categories: high, medium-high, medium-low, and low, based on *Nomenclature Statistique des Activités Économiques dans la Communauté Européenne* (NACE) (Rev.1) and International Standard Industrial Classification (ISIC) (Rev.2). The high-tech class covers two-digit NACE 30, 32, 33, 64, 72, and 73, and three-digit NACE 24.4 and 35.3, which incorporate sectors ranging from computers, communications, aerospace to pharmaceuticals. Eurostat's Science and Technology (Patent Statistics) database provides NACE-based data on the number of patents granted by the USPTO per million inhabitants. For weakly coordinated economies, I aggregate the number of patents across these "cutting-edge" sectors; for strongly coordinated economies, I use patents from "mature" technologies instead, aggregating across NACE digits outside the high-tech class. The OECD Structural Analysis database provides NACE-based data on R&D expenditure. Following the same procedure, I aggregate R&D spending across "cutting-edge" and "mature" technologies respectively for weakly and strongly coordinated economies.

The fine-grained Hicks–Kenworthy coordination index provides the flexibility to slice OECD countries different ways. First, I use data from "cutting-edge" innovation for countries whose Hicks–Kenworthy coordination score is below 0.3, and "mature" innovation for those above. This neatly separates the low-scoring Anglo-Saxon countries from the rest of the OECD, and conforms to the "Europe vs. America" framework that has played an important role in the comparative literature on affluent capitalist economies (Kitschelt et al. 1999; Hall and Soskice 2001; Pontusson 2005; Kenworthy 2004). Second,

I use 0.5 (median) as the threshold, which moves France and Italy from the mature to the cutting-edge group, echoing suggestions in the literature that state intervention undermines interest group coordination and stimulates radical innovation in Southern European capitalism (Amable and Boyer 2001; Trumbull 2004). As a third alternative in slicing varieties of capitalism, I assign all countries scoring above 0.8 (Sweden, Norway, Finland, and Austria) into the cutting-edge innovation group, using data from mature technologies for other remaining countries. Capturing three out of the four main Nordic countries, this 0.8 threshold for cutting-edge innovation is intended to take into account Nordic Europe's high-tech breakthrough (such as in telecommunications) (Kristensen and Lilja 2011), which is referred to as "creative corporatism" by Ornston (2012, 2013). As I show later, my core findings about the effect of coordination on public and insider information are robust to these differing perspectives on who excels in cutting-edge or mature technologies. While splitting innovation into cutting-edge and mature categories addresses the possibility that different countries specialize in different types of innovation, this method also raises its own concern: specialization is relative rather than absolute, and both types of innovation may still be important for all countries. For this reason, I will also test the findings' robustness without distinguishing cutting-edge and mature technologies.

Control Variables

In hypotheses H1 and H2, the financial market phenomenon under examination is adverse selection, where poor information reduces the marginal contribution of equity investment to innovation output, measured by patents. In order to parse out investor uncertainty (i.e. poor insider or public information) from other causes of low innovation yield (such as insufficient R&D input), I start by controlling for R&D expenditure, available from the OECD's Structural Analysis database. Furthermore, public equity may do little to boost innovation output if the firm raises the money for signaling purposes rather than investment. When aggressive legal institutions of shareholder protection confer upon shareholders an active role in enforcing managerial performance, managers may be under strong pressure to raise public equity just to signal managerial talent (or firm value) in the managerial (or mergers and acquisitions) market (O'Sullivan 2000). On the other hand, theories of "managerial entrenchment" suggest an opposite outcome: when shareholders play an active role in disciplining management, incumbent managers may devote themselves to "high-variance" projects (such as technological innovation). This not only deters potential challengers from entering the managerial market but also forces remaining challengers to ask for higher salaries (as risk premium), which weakens their competitive position on the managerial market (Edlin and Stiglitz 1995). Therefore, strong shareholder protection may work both ways: its "signaling effect" may divert managerial attention away from innovation, but its "entrenchment effect" may encourage greater attention and boost innovation output. To control for the strength of legal protection for shareholders, I use the antidirector index developed by La Porta et al. (1998) and extended by Pagano and Volpin (2005), which captures various legal channels empowering shareholders to actively discipline managers, such as proxy by mail, preemptive rights at new equity offerings, and unblocking of shares before shareholder meetings. Through a "participation effect" in the spirit of Goldfarb and Xiao (2011), tough shareholder protection in *public* equity may also affect returns from *private* equity investment, by deterring endogenous entry into the competitive managerial market and worsening the management pool that venture capitalists choose from. For this reason, I include the antidirector index for both public (H1) and private (H2) equity.

I also include center-left share of government seats to consider the possibility that unproductive equity finance may result from party political rather than informational constraints: when social democratic governments prioritize stakeholder over shareholder value, redistributive demands faced by firms may interfere with their optimal use of external finance (Roe 2002). The state may also affect innovation through its role as a core procurer of defense-related technologies such as communication, health, and transportation, with valuable civilian spillovers (Gambardella 1995; Nelson 2005; Freeman and Soete 1997). As a result, I control for defense spending, which deepens both financial and scientific resources for successful innovation. I also control for total government consumption because the state may also facilitate innovation through its general provision of goods and services, many of which are public goods with network externalities or scale economies (such as standardization/certification services) that are complementary to private sector innovation. Data for both defense spending and government consumption are from the Comparative Welfare State Data Set by Huber et al. (2004).

Nonproduction activities (such as trials in controlled environments, development, and design) rely on investment in not only fixed assets but also human capital at the higher (tertiary) level across the economy, beyond the specific sector of innovation (Acemoglu 2003a). To control for a country's general level of investment in human and physical capital, I include both total fixed capital formation as percentage of GDP (OECD's Structural Analysis database, subsection STAN Industry) and gross enrollment ratio in tertiary education (UNESCO Education and Literacy database and World Development Indicators).

Hypothesis H3 examines credit rationing, where poor information reduces the marginal effect of total bank credits on R&D expenditure. Several of the controls above (human capital, fixed assets, and defense spending) are retained, because abundance in these financial/technical resources may not only make R&D more successful but also encourage firms to carry out more R&D. Besides, I also add two additional controls which may affect how much R&D firms choose to carry out. First, as Acemoglu (2003a) and Pissarides (1997) suggest, trade globalization is an important mechanism of technological diffusion, exposing firms to other countries' technologies and forcing them to adapt with indigenous R&D or else lose out to competition. To reflect this possibility, I control for exports plus imports as percentage of GDP, available from the Comparative Welfare State Data Set. Second, the rent-sharing literature (Baldwin 1983; van Reenen 1996) suggests that innovation creates "novelty" rents that may be captured by unions as wage premiums. In rational anticipation of such rent leakage to unions, firms confronted with strong unions may substitute routine production for R&D. In other words, union strength (measured through union members as percentage of wage and salary workers) may discourage firms from investing in R&D.

Results and Discussion

The empirical analyses of hypothesis H1 cover fourteen OECD countries[2] for a period from 1980 to 2002, the latest year in which observations on all control variables are available. I report the output of fixed effect vector decomposition (with Prais–Winsten adjustment for serial correlation) estimations in Table 2.2. I start the discussion with Model 1. In order to ensure that financial markets in different countries are evaluated in the informational environment most relevant to these countries' world of innovation, Model 1 uses innovation data in "mature" technologies for countries scoring above 0.3 on the Hicks–Kenworthy index and in "cutting-edge" technologies for those scoring below. Hypothesis H1 describes what we should observe if coordination reduces the quality of public information: as coordination strengthens, adverse selection in innovation financing worsens, which lowers the marginal contribution of IPOs to innovation output. In other words, the hypothesis predicts an interaction effect between IPOs and coordination, which should be negatively signed.

To interpret the interaction, it is helpful to start with the "main effect" variable for IPOs, because this captures the effect of public equity investment

---

[2] These countries are Australia, Belgium, Canada, Denmark, Finland, France, Germany, Italy, Japan, the Netherlands, Norway, Sweden, the UK, and the US.

Table 2.2. Poor Public Information in IPOs Finance, 1980–2002

	(1) High-tech if HK<0.3[†]	(2) High-tech if HK<0.5	(3) High-tech if HK>0.8	(4) Italy/France dropped	(5) Nordic dropped	(6) Australia dropped	(7) Hall–Gingerich measure of coordination	(8) All patents (high-tech and mature)
Coordination* IPOs	−120.73(11.2)***	−118.12(6.26)***	−116.76(11.1)***	−117.23(12.7)***	−129.71(17.8)***	−139.18(12.9)***	−177.8(10.3)***	−141.36(16.38)***
IPOs	86.51(6.48)***	84.12(6.26)***	79.85(6.21)***	82.93(7.03)***	83.94(7.95)***	102.19(8.34)***	111.44(6.06)***	100.18(11.29)***
R&D	33.48(1.96)***	33.63(1.96)***	33.30(2.25)***	34.60(2.23)***	31.51(4.73)***	33.63(2.08)***	33.04(1.97)***	64.02(2.61)***
Fixed capital	−0.23(0.04)***	−0.22(0.04)***	−0.19(0.03)***	−0.20(0.05)***	−0.26(0.05)***	−0.38(0.04)***	−0.59(0.03)***	−1.39(0.10)***
Tertiary education	0.11(0.06)*	0.10(0.05)*	0.09(0.05)*	0.08(0.07)	0.16(0.07)**	0.02(0.06)	−0.00(0.05)	0.24(0.08)***
Defense spending	2.96(1.48)**	0.369(1.33)***	4.67(1.21)***	3.46(1.62)**	6.88(2.08)***	0.66(2.02)	2.03(1.24)	−12.56(1.18)***
Government consumption	0.57(0.26)**	0.54(0.27)*	0.49(0.24)**	0.70(0.32)**	0.67(0.38)*	0.86(0.31)***	0.56(0.24)**	0.54(0.37)
Minority shareholder protection	0.00(1.17)	0.00(0.96)	0.00(1.02)	0.00(1.62)	0.00(1.8)	0.00(1.9)	0.00(1.13)	0.00(1.10)
Left cabinet	2.13(1.99)	2.78(1.94)	1.88(1.55)	3.29(2.59)	3.45(3.19)	2.74(2.47)	2.75(1.79)	7.07(2.49)***
Coordination	−126.79(14.6)***	−150.04(13.9)***	−89.46(14.85)***	−142.48(16.3)***	−166.85(24.6)***	−84.1(19.19)***	−176.7(13.6)***	−181.1(11.36)***
Constant	64.23(11.68)***	106.23(11.2)***	126.85(12.4)***	83.69(13.16)***	66.78(14.69)***	32.88(15.17)**	127.97(11.0)**	84.91(18.30)***
R-squared	0.99	0.99	0.99	0.99	0.99	0.99	0.99	0.94
N	124	124	131	108	90	111	124	138

* $p<0.1$, ** $p<0.05$, *** $p<0.01$ [†] HK: Hicks–Kenworthy score of coordination

on innovation output when coordination is absent. This will set the stage for the interaction effect that will unfold with the onset of coordination. In Model 1, the "main effect" point estimate for IPOs is 86.51 and significant at the 0.01 level. In other words, when coordination is at its minimum (zero on the Hicks–Kenworthy index), each increase by 1 percent of GDP in IPOs purchase can boost innovation output by around eighty-six patents per million inhabitants, equivalent to about a third (32 percent) of the mean patent output across the countries in the analysis. However, this is very much the upper bound on the effect of public equity investment in stimulating innovation output. For most affluent capitalist economies, the strength of coordination is well above the minimum, and as the negatively signed interaction indicates, the marginal effect of IPOs on innovation output declines as capitalism becomes more coordinated. Based on Model 1's estimate, a one-unit increase in the strength of coordination will reduce the marginal contribution of IPOs to innovation output by around 120 patents per million inhabitants, which is about 45 percent of the average patent output across the countries in the analysis.

To aid the illustration of the interaction, Figure 2.1 plots the marginal effect of IPOs on patent output, conditional on the strength of coordination, with dashed lines marking the 95 percent confidence interval. The direction of the slope reveals the direction of interaction: a downward (upward) slope implies that coordination depresses (enhances) the marginal effect of IPOs on innovation output. The positioning of the confidence interval determines statistical significance: where the confidence interval is above or below zero, the marginal effect is significant, and where the confidence interval encompasses zero in between, the marginal effect is insignificant. As the graph shows clearly, the effect of public equity finance on innovation output diminishes as coordination strengthens. When coordination is minimum at zero, 1 percent of GDP increase in IPOs purchase can boost innovation output by eighty-six patents per million inhabitants. This effect, of course, can also be directly recovered as the "main effect" coefficient for IPOs in Model 1. This impact declines steadily as coordination gathers strength, and drops out of significance after coordination grows to around 0.57 on the Hicks–Kenworthy index (close to the level of the Netherlands).

To better appreciate the substantive implications of such interaction, I follow the procedure by Kam and Franzese (2007) and calculate the marginal effects (and associated standard errors) of IPOs purchase at concrete levels of coordination corresponding to different countries. For example, while 1 percent growth of IPOs purchase in weakly coordinated Australia (0.17 on the Hicks–Kenworthy index) can lift innovation output by sixty-six patents (with standard error at 6.77), the same increment of public equity investment in Japan (0.77 on coordination) has no statistically significant effect on

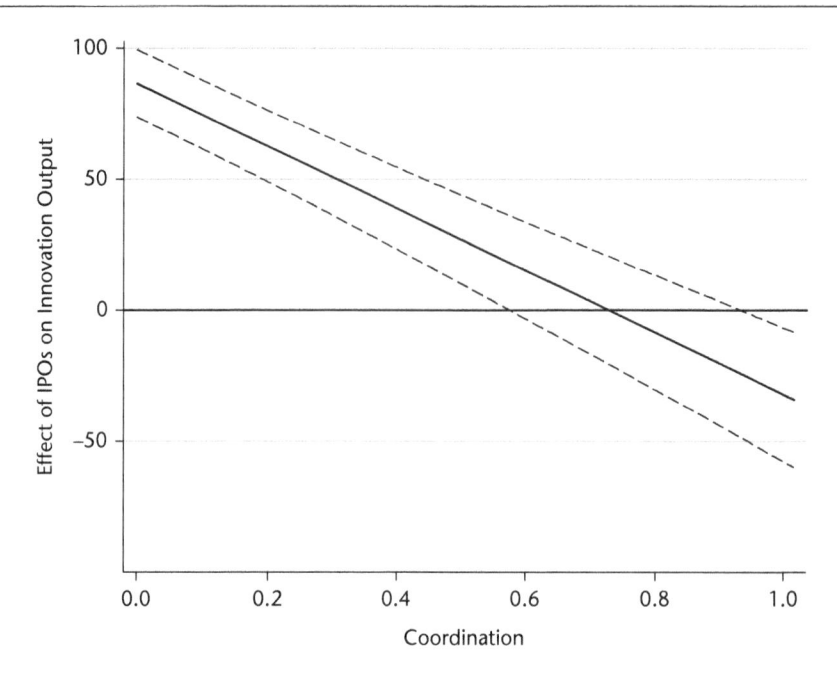

Figure 2.1. IPOs and Innovation

innovation output: the standard error is 10.83, larger than the marginal effect of public equity, which is also wrongly signed at −6.44. In other words, adverse selection in public equity is far more severe in Japan. This difference translates into a shortfall of sixty-six patents in the equity market's performance to attract successful innovation, which is as much as 25.19 percent of the average innovation output across countries. On the other hand, the virtual absence of coordination in the US (0.02 on the Hicks–Kenworthy index) substantially alleviates adverse selection, allowing each 1 percent growth in public equity to generate as many as eighty-four patents, which is almost five times the performance (sixteen patents) in more heavily coordinated Netherlands (0.58 on the Hicks–Kenworthy index). As yet another example of how coordination worsens adverse selection in public equity finance, in the densely organized Austrian economy (0.96), investors actually get a negative return (−29 patents) for each increment of their investment, and this negative effect is significant at the 0.05 level (standard error 12.59). This negative effect of IPOs can also be observed directly in Figure 2.1, where the confidence interval enters negative territory after coordination grows to above 0.90 on the Hicks–Kenworthy index, covering countries such as Austria, Norway, and Sweden. In short, the findings support H1's prediction of what we should observe if coordination suppresses public information: it aggravates adverse selection in public equity, reducing the marginal contribution of equity investment to innovation output.

Model 1 used patent data in "mature" technologies for countries scoring above 0.3 on the Hicks–Kenworthy index and in "cutting-edge" technologies for those scoring below. This corresponds to the "Europe vs. America" comparative framework that has played a key role in the comparative political economy literature (Kitschelt et al. 1999; Hall and Soskice 2001; Pontusson 2005; Kenworthy 2004). Model 2 uses instead the threshold of 0.5, moving Southern European countries into the group of "cutting-edge" innovation. As a third alternative of slicing mature and cutting-edge innovation, Model 3 adopts a threshold of 0.8, assigning all countries above it (three Nordic countries plus Austria) to the cutting-edge innovation group, in order to give greater emphasis to recent Nordic success in radical innovation as highlighted by Ornston (2012, 2013).

Besides different ways of slicing innovation, there are also different approaches to classifying coordination. Of particular relevance to this chapter's insider/public information tradeoff are the Southern European countries that mix symptoms of strong and weak coordination. The ambiguous coordination capacity of these hybrid countries is evident in their middle-of-the-road Hicks–Kenworthy scores (0.44 for Italy and 0.40 for France). The inclusion of these countries may have suppressed the potential interaction effect of coordination in two ways. First, as Hall and Gingerich (2009) point out, hybrid regimes may be deprived of the institutional synergies from both strong and weak coordination, and consequently perform worse than both. In other words, coordination in these countries may be both too strong (to boost public information) and too weak (to boost insider information), in effect severing the link between coordination and information quality. Second, in these countries, active state intervention in the banking sector helps to reduce lending risks through various mechanisms, such as lender guarantors and interest rate subsidies (Loriaux 1991). This will diminish the weight of information opacity in investor strategies, and hence limit the extent to which the impact of coordination on information quality may be observed in the lending market. Although this second concern may be of particular relevance to only H3 (credit rationing), the first concern affects all three hypotheses. In Model 4, therefore, I re-estimate Model 1 by dropping France and Italy, to check whether the interaction effect is indeed notably larger after these two possible "no effect" countries are removed from the analysis. As Table 2.2 shows, Model 4's interaction coefficient is no larger than Model 1.

Along the same vein, Models 5 and 6 consider two other important "subtypes" in varieties of capitalism. Although most continental and Northern European countries feature relatively strong coordination, labor's political dominance and the public sector's economic weight set Nordic countries apart from continental Europe in the patterns of coordination (Pontusson 2005; Amable 2003). Furthermore, this Nordic/continental distinction in coordination directly affects patterns of innovation: unlike their continental counterparts, several Nordic economies have become breakthrough players in cutting-edge technologies (such as telecommunications) (Ornston 2012; Kristensen and Lilja 2011). In Model 5, I drop Nordic countries from the analysis to see if this subtype substantively affected the core findings. Model 6, in turn, considers the Antipodean subtype of Anglo-Saxon economies by dropping Australia from the analysis.

Although these additional models differ somewhat in how they slice the world of innovation or varieties of capitalism, their findings are largely similar to Model 1: the point estimate for the interaction is always significant and signed correctly (negative), consistent with H1's prediction of adverse selection in public equity when coordination is strong. Because there are no substantial differences in findings, I omit separate visual display of interaction for these five remaining models. In Model 7, I re-estimate Model 1, but this time with coordination measured through the Hall–Gingerich index. As noted earlier, although Hall–Gingerich shares with Hicks–Kenworthy a similar conceptual approach to coordination (with firms and investors as focal players), it places relatively more emphasis on the labor side of coordination, adopts a different computational approach (deriving country scores from factor analysis loadings), and ranks various countries somewhat differently from Hicks–Kenworthy (such as the relative position of Germany vis-à-vis Nordic Europe, the placement of Italy and Belgium, as well as the position of Ireland, as seen in Table 2.1). In other words, the Hall–Gingerich index leaves sufficient distance from Hicks–Kenworthy to make it a suitable check for the robustness of findings. As Model 7 shows, the core pattern of finding is not substantively altered by the switch from Hicks–Kenworthy to Hall–Gingerich. The point estimate for the interaction between IPOs and coordination continues to be significant at the 0.01 level and negatively signed, and the magnitude of the interaction is also not substantially different from earlier ones. Based on Model 7, Figure 2.2 plots the marginal effect of IPOs conditional on coordination.

Comparing Figure 2.2 with the earlier Figure 2.1 (which measured coordination through Hicks–Kenworthy), the pattern of interaction from these two graphs is very similar, other than some difference in the precision of estimation (i.e. width of the confidence interval), and slight difference in where the interval starts to cross from positive into negative territory. In other words, regardless of whether it is measured from Hall and Gingerich's data or Hicks and Kenworthy's data, there is evidence that coordination reduces the marginal effect of public equity investment on innovation output. This evidence

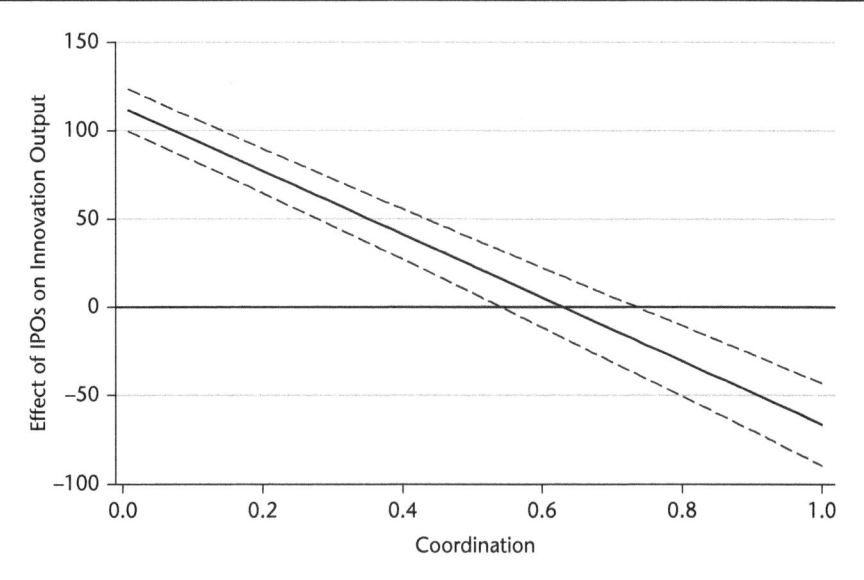

Figure 2.2. IPOs and Innovation

is consistent with H1's prediction of what we should observe if coordination suppresses public information.

Because different countries may specialize in different types of innovation, preceding models split innovation into high-tech and mature categories. However, in reality, despite relative differences in specialization, countries still engage in both types of innovation. Therefore, as an additional robustness check, I also examine the impact of IPOs on total innovation output, without distinguishing between high-tech and mature categories (Model 8). Again, the core pattern of finding (negative interaction between coordination and IPOs) is unchanged.

The next two hypotheses describe what we should observe if coordination increases the quality of insider information: as coordination strengthens, R&D financing should experience less adverse selection in private equity (H2), and less credit rationing in lending (H3). I start with H2. Venture capital financing is very strongly tied to high-risk cutting-edge innovation. For example, according to the 2012 Yearbook of the National Venture Capital Association (Thompson Reuters 2012, p. 53), $8,103 million out of 2011's total $9,922 million venture-backed IPOs went to firms closely related to high-tech innovation, such as computers, IT, biotechnology, software, telecommunications, and media/entertainment. Therefore, it may be especially relevant to study the impact of venture capital investment on "cutting-edge" technologies. To do so, I use innovation data from "cutting-edge" technologies for *all* countries, rather than "cutting-edge" for some and "mature" technologies for others. Although high-risk high-tech innovation captures the lion's share of venture capital investment, a substantial amount, in absolute terms, still goes to other, more mature, sectors. For this reason, later in the analysis I will check robustness by examining all countries on the basis of mature technologies as well.

Pooled times series data for venture capital investment is unfortunately limited, starting only from 1997. As a result, Table 2.3a's analysis of venture capital covers a relatively short six-year period from 1997 to 2002. In order to preserve sufficient degrees of freedom in analyzing this short time period, I exclude from estimation the variable on tertiary education, which has very sporadic data coverage.

H2 describes what should occur if coordination improves insider information: as coordination strengthens, private equity should experience less adverse selection, and as a result the marginal contribution of venture capital to innovation output should increase. Model 1 in Table 2.3a bears this pattern out. Again, it is helpful to start the discussion with the "main effect" coefficient for venture capital, which captures the effect of financial markets when coordination is absent. This establishes the bottom line against which to assess the performance of financial markets when there is stronger coordination.

As Model 1 in Table 2.3a shows, when there is minimum coordination (zero on the Hicks–Kenworthy index), venture capital investment has little impact in boosting high-tech innovation output. Each increase by 1 percent of GDP in venture capital investment translates into no more than two patents per million inhabitants, with a standard error more than twice as large (4.9). This, nevertheless, is only the *worst* possible outcome for venture capital investment. The positive sign of the interaction in Model 1 suggests that the ability of venture capital to stimulate innovation output will grow stronger as coordination becomes denser. A one-unit increase in coordination will raise the marginal contribution of venture capital investment to innovation output by around forty-three patents per million inhabitants, which is about 17 percent of the average patent output across the countries in the analysis. In other words, compared with minimum coordination, a country with average coordination (0.51 on the Hicks–Kenworthy index) will add about twenty-one patents to the economy (equivalent to 8 percent of the average innovation output across countries) for each 1 percent of GDP increment in venture capital investment.

To illustrate this pattern of interaction, Figure 2.3a plots the marginal effect of venture capital on patent output, conditional on the strength of coordination. As the graph shows clearly, the effect of venture capital on innovation output strengthens as coordination grows. In the leftmost region of the slope (less than 0.2 on the Hicks–Kenworthy index), the confidence interval straddles zero, implying that the effect of venture capital on high-tech innovation output is indistinguishable from zero when coordination is very weak. This, of course, is consistent with the statistically insignificant "main effect"

Table 2.3a. Good Insider Information in Venture Capital Finance, 1997–2002

	(1)	(2)	(3)	(4)	(5)	(6)	(7)	(8)
	High-tech (all countries)	US dropped	UK dropped	US/UK/Aus dropped	Italy/France dropped	Nordic dropped	Mature tech (all countries)	Hall–Gingerich measure of coordination
Coordination*	42.8(12.2)***	31.0(13.9)**	44.9(14.1)***	61.6(22.1)***	62.5(15.4)***	54.5(15.6)***	66.81(10.91)***	18.88(7.87)**
Venture capital								
Venture capital	1.99(4.9)	6.58(5.47)	1.56(5.22)	−6.88(10.54)	2.67(5.52)	5.35(5.32)	−32.94(4.81)***	5.71(4.39)
R&D	16.53(2.5)***	18.32(2.6)***	15.99(2.7)***	18.55(2.8)***	12.51(2.7)***	10.94(3.2)***	−2.14(1.37)	19.55(2.56)***
Fixed capital	−0.84(.05)***	−1.01(.07)***	−0.82(.06)***	−1.57(.42)***	−0.92(.07)***	−1.03(.08)***	−1.86(.05)***	−0.75(.05)***
Defense spending	4.01(1.16)***	2.83(1.62)	3.41(1.49)**	7.31(1.76)***	6.12(1.72)***	9.11(2.02)***	1.28(1.37)***	3.38(1.22)**
Government consumption	−0.29(0.23)	−0.23(0.25)	−0.32(0.24)	−0.31(0.26)	−0.49(0.25)*	−0.36(0.33)	−0.02(0.22)	−0.29(0.24)
Minority shareholder protection	0.00(0.64)	0.00(0.67)	0.00(0.72)	0.00(0.69)	0.00(0.86)	0.00(1.39)	0.00(0.47)	0.00(0.64)
Left cabinet	−5.91(2.32)**	−6.34(2.4)**	−6.04(2.64)**	−7.89(2.7)***	−6.7(2.94)**	−8.27(3.01)**	−7.16(1.91)***	−5.46(2.41)**
Coordination	−33.8(3.3)***	−70.6(5.1)***	−32.8(4.2)***	−40.3(4.2)***	−31.2(4.4)***	−41.4(7.8)***	−21.62(2.83)***	−27.19(2.33)***
Constant	129.8(3.9)***	159.7(3.9)***	130.9(4.5)***	140.1(9.7)***	133.3(5.1)***	134.6(5.3)***	303.62(3.92)***	121.44(4.1)***
R-squared	0.97	0.95	0.98	0.96	0.97	0.98	0.99	0.98
N	50	47	46	39	42	39	51	50

* $p<0.1$, ** $p<0.05$, *** $p<0.01$

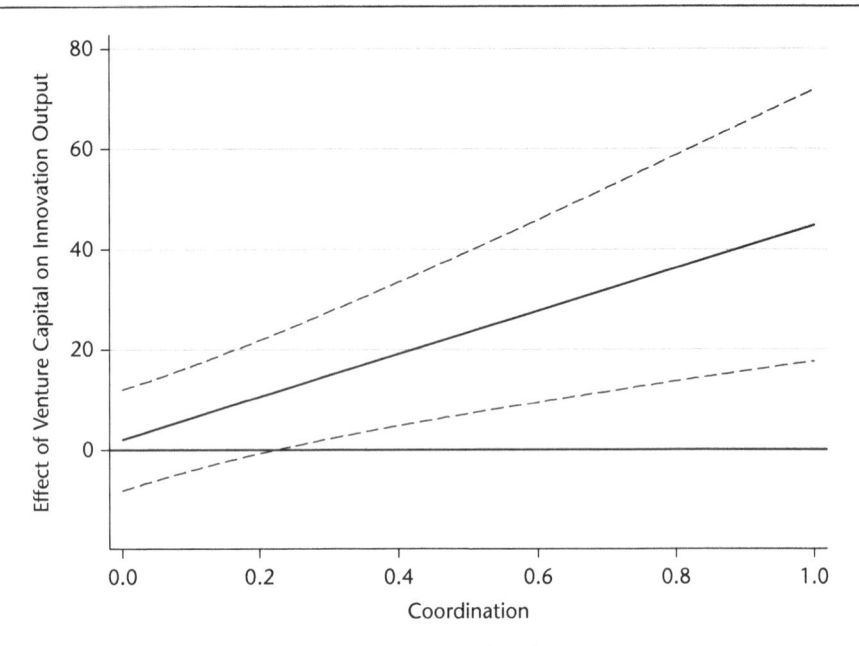

Figure 2.3a. Venture Capital and Innovation

coefficient for venture capital in Model 1. As the economy becomes more coordinated, the boost to innovation output becomes progressively larger: at median coordination (0.5 on the Hicks–Kenworthy index, in-between Switzerland and Italy), each increment of venture capital injection can lift innovation output by twenty-three patents, and by the point of maximum coordination, the same increment of venture capital is sufficient to add forty-five patents, more than twenty times the impact when coordination is absent.

The leftmost region of the graph, where the confidence interval dips below zero, covers the entire set of Anglo-Saxon countries, all of which score below 0.2 on the Hicks–Kenworthy index. In other words, venture capital fails to have any effect in lifting high-tech innovation output precisely in these countries. This finding is particularly striking because, as noted earlier, some of these countries (such as the US and the UK) are the very markets that attract large amounts of venture capital for the high-powered risk premium it offers (OECD 1995; Lamoreaux and Sokoloff 2007). As a result, poor insider information would have posed the least hindrance in such countries, making them a "hard test" for the chapter's theory that insider information is poorer in less coordinated economies.

In Models 2 and 3 respectively, I rerun the analysis by dropping the US and the UK, both of which are venture capital outliers as percentage of GDP (respectively two and 4.5 times the mean size of venture capital across countries in the analysis). In Model 4, I drop three countries together: the UK, the US, plus Australia (another outlier with venture capital 1.8 times the cross-country average). With thirty-nine observations left and ten parameters (including an interaction), this reduces the degrees of freedom close to the limit. If the benefits from generous risk premium in these outliers really neutralized the impact of poor information, then when these countries are dropped the intensity of interaction between coordination and venture capital should rise notably. As Models 2 through 4 show, however, dropping these countries does not notably sharpen the interaction effect in a consistent way. This suggests that the interaction effect is not necessarily weaker even in these very venture-friendly environments. In Models 5 and 6 respectively, I dropped two subcategories of capitalism (Southern European and Nordic), and the results are not qualitatively different (and therefore separate graphs omitted). While preceding models focused on high-tech innovation, Model 7 further checks robustness by estimating the innovation outcome in mature technologies instead. Finally, as one additional robustness check, I reanalyze Model 1 by substituting with the Hall–Gingerich measure of coordination in Model 8. The core pattern of finding remains unchanged: the interaction between coordination and venture capital is positively signed, implying that coordination boosts the impact of venture capital investment on innovation output.

Having established a core pattern of finding that venture capital is, counterintuitively, more effective in more densely coordinated European economies, I now further consider some additional implications that may flow from this core finding. In particular, I address the following three questions: (1) Is the European advantage more evident in the "early" or "late" stage of venture capital investment? (2) Did the recent systemic shock to global financial markets sharpen or mask the contrast in venture capital performance between liberal market and strategically coordinated economies? (3) Was the inferior performance of Anglo-Saxon venture capital simply the result of diminishing returns? Although all three are very interesting inquiries that further shed light on venture capital, data that can be leveraged for these analyses is very limited. As a result, in contrast to the preceding core analyses in Table 2.3a, the following analysis involves more improvisation in data construction, much smaller numbers of observations, and hence necessarily more speculative (as opposed to conclusive) findings.

I start with the question of "stages" in venture capital. The finding of stronger venture capital performance in Europe than Anglo-Saxon countries is driven by a theory of information: strategically coordinated capitalism is more effective than liberal market capitalism in communicating insider information, which venture capitalists rely on to screen and monitor ongoing projects. Venture investors' need for insider information may be especially acute at early stages of investment (such as seed, startup, and first-round), where the vested firm focuses on forming ideas, building business plans, and initial product development and design. Because there is yet no commercial production or sale to "test the project," venture capitalists face significant uncertainty. After the project has survived for several years and commercial production has started, venture capital enters its later stages (such as growth, second-round, and mezzanine). By this point, there will be more "hard information" from sales and revenue, which gives investors more publicly available data to assess the project, reducing the weight of insider information. Therefore, if insider information quality is at the heart of the contrasting performance between European and Anglo-Saxon venture capitalists, this contrast may be stronger where insider information has more bite: in other words, the interaction effect between coordination and venture capital should be sharper in early than later stages of investment.

However, venture capital data by investment stage is not only very short across time, but also inconsistent across countries, especially for early-stage investment. Eurostat's Science and Technology Database measures two important substages of early-stage venture capital (seed and startup) for European countries, but has no data on first-round financing. On the other hand, some of the early-stage venture capital data for Anglo-Saxon countries (gleaned from yearbooks and annual reports of Canadian, American, and Australian national venture capital associations) either does not distinguish between seed and startup, or roll startup and first-round into one aggregate statistic. For this reason, instead of further disaggregating my measure into these finer substages, I simply calculate a measure of total early-stage investment for each country, by summing over all substages with available data. I follow the same procedure and calculate each country's total later-stage venture capital investment. Because the data is extremely short (from 2007 to 2012), to preserve sufficient degrees of freedom I drop R&D expenditure from the list of controls, since its coverage goes only to 2010 or 2011 for some countries (and with gaps), cutting off the precious data points available on the venture capital measure. Even with this step, it is important to acknowledge that most of the following analyses have very low degrees of freedom, relying on as few as thirty-seven observations.

I re-estimate Model 1 of Table 2.3a, this time respectively for early-stage and later-stage investment, and report the results in Table 2.3b (Models 1 and 2 respectively). If it is the quality of insider information that drives the outcome, then the interaction between coordination and investment should be stronger where insider information is more important (i.e. early-stage investment). A comparison of Models 1 and 2 in Table 2.3b indeed bears this out: the interaction coefficient for early-stage investment is more than twice the estimate for later-stage investment. This implies that each increment of transition

Table 2.3b. Good Insider Information in Venture Capital Finance, Additional Analysis

	(1) Early Stage (Seed and Startup)	(2) Later Stage	(3) Buyout	(4) Before Onset of Crisis	(5) After Onset of Crisis	(6) Controlling for New Firm Entry
Coordination* Venture capital	**162.83(38.41)*****	**78.81(24.18)*****	**24.17(32.89)**	**49.34(20.12)****	**83.74(14.53)*****	**348.16(121.64)*****
Venture capital	4.35(23.73)	14.45(10.25)	7.41(6.79)	−2.24(10.08)	4.57(7.92)	−309.49(195.51)
Fixed capital	0.68(0.52)	0.36(0.47)	0.57(0.50)	0.02(0.08)	0.51(0.49)	0.33(0.16)*
Defense spending	43.65(2.89)***	47.81(2.66)***	62.77(2.66)***	45.32(2.79)***	40.88(2.71)***	−122.42(23.31)***
Government consumption	−1.68(0.37)***	−1.17(0.35)***	−2.56(0.38)***	1.10(0.54)*	−.46(0.36)	15.92(2.44)***
Minority shareholder protection	0.00(1.61)	0.00(1.10)	0.00(1.16)	0.00(1.34)	0.00(1.12)	27.89(6.42)***
Left cabinet	10.39(4.84)***	8.42(4.42)*	9.09(5.43)	−11.55(5.00)**	8.84(4.48)	27.78(19.02)
Coordination	194.86(7.10)***	217.06(5.53)***	236.81(5.76)***	110.01(7.39)***	193.24(6.04)***	−147.04(56.41)**
Enterprise birth rates	n.a.	n.a.	n.a.	n.a.	n.a.	−13.19(5.57)**
Enterprise birth rates* Venture Capital	n.a.	n.a.	n.a.	n.a.	n.a.	17.09(12.98)
Constant	7.10(16.38)	−10.47(15.21)	−12.92(16.42)	148.56(9.78)***	−10.29(15.07)	−147.92(97.74)
R-squared	0.99	0.99	0.98	0.97	0.99	0.87
N	65	66	66	128	65	37

* $p<0.1$; ** $p<0.05$; *** $p<0.01$

from liberal market to strategic coordination in the economy can improve early-stage venture capital performance twice as fast as later-stage investment. In other words, differences in coordination strength matter most precisely where insider information is most valuable. While the contrast between the two investment stages highlights the role of insider information, evidence from each stage on its own also echoes the earlier core finding: the interaction is signed correctly (positive), and comfortably clears significance at the 0.01 level.

Because differences in coordination strength matter more in stages where insider information is more valuable, I create an additional placebo test by examining the impact of coordination on investment *after* venture capital, i.e. "buyout" investment, whereby assets and liabilities of the previously venture-backed firm are acquired by other private equity investors (in case the firm has already gone public, this involves the return to private equity through de-listing). By this point, the firm has undergone all stages of venture capital investment as well as expansion in production and sales, all of which generate additional information for buyout investors. As a result, the firm's hidden insider information may be less pivotal to buyout investors than venture capital investors. Furthermore, if the firm is previously publicly listed, knowledge about its performance will also be available in the form of public information released during the IPOs process. Since insider information may be mixed with public information during buyout investment, the impact of good insider information in strategically coordinated economies may be especially negligible. In other words, compared with both stages of venture capital, the interaction between coordination and investment may be especially weak during buyout. This turns out to be indeed the case in Model 3 of Table 2.3b, where I analyze buyout rather than venture capital investment. Although the point estimate for the interaction is still signed positively, it fails to clear statistical significance at the 0.1 level. In other words, differences in the quality of insider information (driven by differences in coordination strength) do not affect performance in buyout investment. Overall, the stepwise decline in the interaction effect as we move from early- to later-stage and then to buyout investment is consistent with my core argument that the quality of insider information drives the (counterintuitive) European advantage over Anglo-Saxon economies in venture capital performance.

How may the recent global financial crisis affect our ability to observe the European advantage in venture capital performance? There are differing perspectives on how this large negative shock to financial markets has originated. Some focus on *adverse conditions* (such as capital mobility, growing competition and fungibility among financial products, faster transactions, and closer to this chapter's core theme, lack of information); others focus on *ill incentives* (such as excessively high-powered compensation for executives and moral hazard for banks being rescued by governments) (Tafara 2012). Although these two perspectives on the crisis are closely connected (for example, fierce competition may exacerbate the distortion of incentives), they do have different implications for how the crisis may affect the performance of financial markets. While there are always wrongly invested "toxic" projects with poor prospects, a systemic negative shock will turn more of these poor prospects into reality than normal times, generating widespread failure. The larger the shock, the deeper the scoop among the pool of toxic projects. If poor investment decisions are driven by excessively steep and myopic executive incentives, and if we assume (reasonably) that executives everywhere act on (rather than resist) such incentives, then a negative shock should hit all countries to a similar extent, allowing the crisis to *mask* the contrast between European and Anglo-Saxon venture capital performance. By contrast, if poor investment decisions are caused by the lack of insider information, the pool of toxic projects may be larger where such information is more limited (Anglo-Saxon economies). As a result, a systemic negative shock may scoop out more failures in these economies as well, allowing the crisis to further *sharpen* the contrast in performance between liberal market and coordinated economies.

Unfortunately, because of the very recent onset of the global financial crisis, the post-onset data series on venture capital and innovation are too short for very rigorous quantitative assessments of the conjectures above. Instead, I adopt a simpler approach far less demanding on the data: I run two separate estimations, respectively for the time period before the onset of the crisis (before 2007) and afterwards (after 2007). For most countries (i.e. those measured by the Eurostat's Science and Technology Database), venture capital data before 2007 is only available as total investment, making no distinction between various stages. To make comparison across time possible, I also derive a measure of total investment for the post-2007 period, summing over early- and later-stage investments. Because, as noted earlier, the data by investment stage is incomplete in its conceptual coverage (for example, missing seed investment for some countries and first-round investment for others), the sum over the stages will also be an incomplete description of total venture capital investment after 2007. This will no doubt affect the comparison with the pre-2007 period, and for this reason the finding should be regarded as no more than speculative.

In Models 4 and 5 of Table 2.3b respectively, I estimate the effect of venture capital investment for the precrisis and crisis period. If the toxic projects exposed by the crisis were caused by "ill incentives," all countries should be similarly affected, severing the link between outcome and coordination. In other words, the coordination–investment interaction should be *weaker* for the crisis period than the precrisis period. By contrast, if the toxic projects were the result of poor insider information, the impact of the crisis will be larger where such information is poorer. In other words, the coordination–investment interaction should be *stronger* in the crisis period. A comparison of the interaction coefficients in Models 4 and 5 appear to favor the latter interpretation of the crisis: the interaction coefficient in the crisis period is larger than the precrisis period, and also clears significance more comfortably. However, unlike the earlier comparison between early- and later-stage venture capital (where one interaction effect is more than twice the other), here the difference between the two interaction effects is less drastic. For this reason, it may be easier to assess the difference graphically.

In Figure 2.3b, I plot the marginal effect of venture capital on innovation output, conditional on coordination, respectively for the precrisis period (solid slope) and crisis period (dashed slope). Both slopes are upward, indicating that both periods support hypothesis H2's core message: venture capital performance is better where insider information is richer (i.e. more coordinated economies). Furthermore, the crisis-period slope is indeed somewhat steeper than the precrisis slope, indicating that the interaction effect is stronger after the crisis onset, an outcome consistent with a "poor information," rather than "poor incentive," interpretation of the crisis. Nevertheless, it is important not to read too much into these findings, given the very small number of observations and the imperfect construction of data used in the analysis.

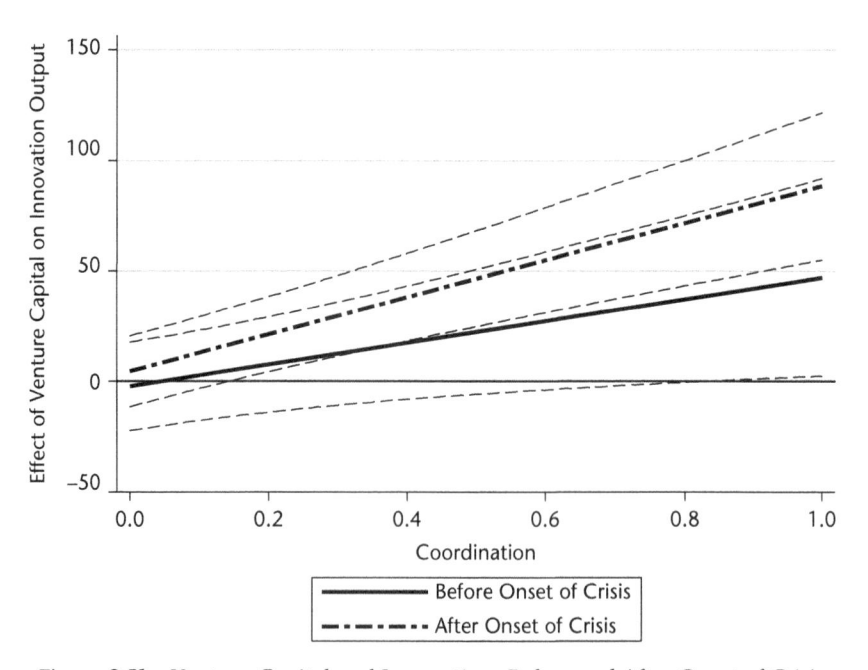

Figure 2.3b. Venture Capital and Innovation, Before and After Onset of Crisis

Finally, there may be another explanation for the surprising outcome of worse venture capital performance in Anglo-Saxon economies: diminishing returns. When there are already many venture investors (e.g. the mature venture capital market in the US), many of the "best picks" among the finite pool of potential innovation projects would have already been located. As a result, the marginal quality of an additional pick by an additional investor is unlikely to be as high. Conversely, where venture capitalists are far and few between (e.g. some small continental European economies), much of the project pool remains untapped, so one additional investor may have a large marginal impact in scooping out a high-quality project. Regrettably, it is difficult to quantitatively assess this alternative explanation, because while data on venture capital by aggregate expenditure are available across countries and years, there are no similar data by the number of investors or projects. Nevertheless, it is important to acknowledge that this potential confounding factor may be an important reason why Anglo-Saxon countries are characterized by a combination of *large* venture capital markets and *poor* venture capital performance.

While the preceding argument focuses on the maturity of *venture capital markets*, we may also rationalize the diminishing returns thesis by, alternatively, focusing on the maturity of *firms*. While venture capital may be crucial in helping young startup firms overcome the large fixed setup cost of R&D, once the project "takes off," the marginal utility from additional private equity injection may decline. In other words, venture capital investors in Europe may perform better simply because they face more firms that are genuinely fresh and untested new entrants. To measure the entry of new firms, I use enterprise birth rates (enterprise birth as percentage of active enterprises with at least one employee) from the OECD's Business Demography Indicators. If the difference in performance is mainly due to difference in the extent of new firm entry, once I control for the interaction between investment and new firm entry, the interaction between investment and coordination should no longer be significant in determining innovation outcome. Unfortunately, this is where the analysis hits the most severe data constraint: data on enterprise birth rates is only available for some occasional years during the 1998 to 2006 period, covering only eight countries with fewer than seven observations each. As a result, the analysis in Model 6 (Table 2.3b), in which I control for the interaction between new firm entry and investment, has only thirty-seven observations. This, of course, pushes the degrees of freedom to the limit, for a model that has to estimate *two* interaction parameters along with *eight* other stand-alone parameters. For this reason, not much should be made of the *numerical value* of the interaction coefficients, and we can probably learn no more than whether the interactions are *signed* in a direction consistent with theoretical expectations. As Model 6 shows, even after the potential interaction with new firm entry is taken into consideration, the effect of venture capital investment on innovation output continues to be stronger in economies with denser insider information, as reflected in the positive interaction between investment and coordination. In other words, the difference across countries in venture capital performance cannot be attributed solely to the presence of diminishing marginal returns to investment in more established firms. Furthermore, Model 6 also provides only very weak evidence for such diminishing returns: although the interaction between venture capital investment and enterprise birth rates (last parameter in Model 6) has a correct, positive, sign (investment has more bite when there are more new entrant firms), it fails to clear statistical significance.

Overall, I have examined three additional implications from the core counterintuitive finding that venture finance, as a key source of risky capital, performs better in coordinated rather than liberal market economies. These three additional inquiries are necessarily no more than speculative, given the extremely low degrees of freedom used to draw my statistical inferences. This acknowledgment aside, all three lines of inquiry led to findings that are consistent with the chapter's core logic that it is the (better) quality of insider information in European economies that drives the (better) performance of their venture capitalists. For example, the cross-country contrast in venture capital performance is largest (smallest) at the investment stage where insider information is most (least) crucial: early stage (buyout). The systemic negative shock of the global financial crisis revealed more "dud" investment projects where insider information is more lacking, sharpening the difference across countries in venture capital performance. Furthermore, even after controlling for possible diminishing returns to investment in more established firms, venture capital continues to perform worse in economies with poorer insider information. Nevertheless, it is important to reiterate that there is another potential channel of diminishing returns which I could not control, due to data limitation: the number of venture capital investors.

After having examined venture capital in great detail, now I turn to another important financial market: bank lending. As examples of respectively "debt" and "equity," lending and venture capital cannot be more *different* from each other, but they are *similar* in both relying heavily on insider information. If the chapter's argument about insider information is correct, we shall continue to see a pattern of more effective investment in coordinated economies, despite all the inherent differences between banking and venture capital. According to hypothesis H3, if coordination indeed increases the quality of insider information, then, as coordination strengthens, credit rationing against innovation projects should be relaxed, and as a result the effect of total bank credits on R&D expenditure should increase. In Model 1 of Table 2.4, the coefficient for the coordination–banking interaction is statistically significant, and correctly signed (positive). A one-unit increase in the strength of coordination will increase the marginal contribution of bank credits to R&D expenditure by 0.72 percent of GDP. In other words, compared with minimum coordination, a country with average coordination (0.51 on the Hicks–Kenworthy index) will add $0.36 of R&D spending to the economy for each dollar of credit provided in the banking sector.

Figure 2.4 plots the marginal effect of bank credits on R&D spending, conditional on the strength of coordination. As the graph shows clearly, the effect of bank lending on firms' R&D expenditure strengthens as the economy becomes more coordinated. When the strength of coordination is below median, the effect of bank lending on R&D spending is mostly indistinguishable from zero, and in regions close to minimum coordination, the impact of banks on enterprise R&D expenditure is actually negative. In other words, when the economy is very weakly coordinated, there is evidence of severe credit rationing against R&D, implying very poor quality of insider information. However, the severity of credit rationing declines as coordination gathers strength. The effect of bank lending starts to have a statistically significant and positive contribution to R&D after coordination grows to around 0.70 on the Hicks–Kenworthy index (close to the level of Denmark). At this level of coordination, each 1 percent of GDP in bank lending is sufficient to raise R&D by 0.22 percent of GDP, and where coordination is closer to maximum strength (Austria, Sweden, and Norway), the same increment of bank lending is sufficient to add 0.42 percent of GDP to R&D expenditure, almost twice the power in Denmark. This pattern of finding supports H3's prediction of what we should observe if coordination boosts insider information: it alleviates credit rationing against R&D, sharpening the contribution of bank lending to R&D expenditure.

The rest of Table 2.4 checks the robustness of findings in a similar fashion to Table 2.2. Models 2 and 3 use different thresholds in determining which countries' R&D data are taken from "mature" or "cutting-edge" technologies. Models 4 through 6 drop various distinct subcategories of capitalism (Southern European, Nordic, and Antipodean) to see if their presence in the analysis significantly altered the core findings. In Model 7, I again re-estimate Model 1 by substituting the Hall–Gingerich index for the Hicks–Kenworthy index. In Model 8, I estimate the impact of bank lending on all R&D, without distinguishing between "cutting-edge" and "mature" technologies for different countries. As the findings from these seven alternative models show, the core pattern remains very consistent: the interaction between coordination and bank lending is positive, significant at the 0.01 level, and relatively similar in magnitude across the models. In other words, various tests in the credit market provide relatively robust evidence for good insider information in coordinated capitalism: less rationing of credit against R&D.

Table 2.4. Good Insider Information in Bank Finance, 1980–2002

	(1)	(2)	(3)	(4)	(5)	(6)	(7)	(8)
	High-tech if HK<0.3	High-tech if HK<0.5	High-tech if HK>0.8	Italy/France dropped	Nordic dropped	Australia dropped	Hall–Gingerich measure of coordination	All R&D (high-tech and mature)
Coordination* Bank credits	0.72(0.12)***	0.94(0.17)***	0.72(0.10)***	0.79(0.12)***	0.50(0.14)***	0.85(0.13)***	0.55(0.13)***	0.75(0.11)***
Bank credits	−0.28(0.08)***	−0.47(0.07)***	−0.23(0.06)***	−0.32(0.09)***	−0.19(0.08)**	−0.39(0.08)***	−0.12(0.08)	−0.18(0.07)**
Fixed capital	−0.003(0.000)***	−0.004(0.000)***	0.000(0.000)	−0.004(0.000)***	−0.008(0.001)***	−0.004(0.000)***	−0.002(0.000)***	0.000(0.000)
Tertiary education	0.007(0.000)***	0.007(0.000)***	0.008(0.001)***	0.007(0.001)***	0.009(0.001)***	0.009(0.001)***	0.009(0.001)***	0.01(0.00)***
Defense spending	0.10(0.01)***	0.08(0.01)***	0.10(0.01)***	0.09(0.01)***	0.13(0.01)***	0.10(0.01)***	0.12(0.1)***	0.13(0.01)***
Trade openness	0.007(0.000)***	0.007(0.000)***	0.008(0.000)***	0.008(0.001)***	0.01(0.001)***	0.008(0.001)***	0.006(0.000)***	0.007(0.000)***
Union density	0.02(0.00)***	0.01(0.00)***	−0.005(0.001)***	0.02(0.00)***	0.04(0.002)***	0.02(0.001)***	0.02(0.000)***	0.003(0.001)***
Coordination	−0.50(0.11)***	−0.43(0.11)***	−1.07(0.09)***	−0.52(0.12)***	−0.43(0.16)***	−0.60(0.12)***	−0.09(0.13)	−0.43(0.10)***
Constant	−0.87(0.09)***	−0.43(0.09)***	0.42(0.09)***	−0.94(0.13)***	−1.59(0.11)***	−0.63(0.12)***	−1.3(0.12)***	−0.21(0.08)**
R-squared	0.94	0.94	0.96	0.94	0.96	0.92	0.94	0.92
N	169	183	178	153	127	154	169	197

* $p<0.1$, ** $p<0.05$, *** $p<0.01$

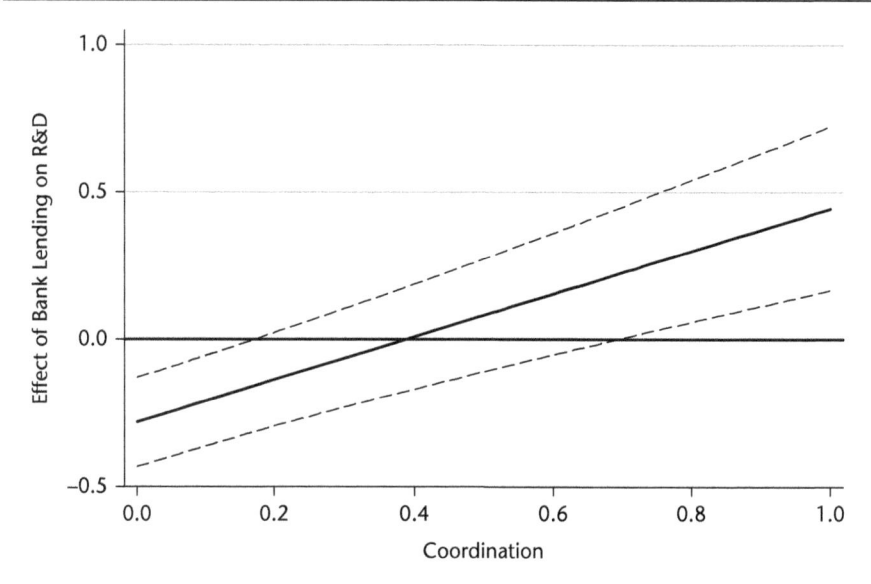

Figure 2.4. Bank Lending and Innovation

The core quantitative analyses in this chapter have provided a broad picture of how nations differ in the performance of their financial markets. While liberal market Anglo-Saxon economies do well in markets that rely on public information (IPOs), strategically coordinated European economies do better where insider information is more important (banking and venture capital). The finding on venture capital is perhaps particularly counterintuitive, given the common association of venture capital with liberal market economies. Here, I supplement the quantitative finding with some additional case study evidence on venture capital from the secondary literature. These brief case studies provide concrete examples of European success in venture capital, as well as the inadequacies of venture capital for innovation financing in the US.

Although venture capital is commonly understood to be important for small high-tech startup firms, Keller and Block (2013) found that, in the US economy, the success of these small firms was driven instead by the state, in the form of a relatively little-publicized government program called the Small Business Innovation Research program (SBIR). Established in 1982 on the passing of the Small Business Innovation Development Act, SBIR has been in existence continuously for the past three decades, precisely the period during which high-tech innovation in the US took off. While the authors identified multiple mechanisms through which SBIR contributed to small firm innovation, I only highlight the financing mechanism, which is the main theme of this chapter. SBIR itself is a small program, disbursing only around $2 billion annually to small innovative firms. The contribution of SBIR to innovation financing, therefore, is not so much in the *funding* it offers, but in the *information* it provides to investors and firms, so that they can match more easily. As the authors pointed out, without the screening and reviewing functions of SBIR, it would have been far more difficult for venture capitalists in the US to locate the right firms. The reliance by venture capitalists on the state to vet potential innovation projects underlines the poor quality of insider information they possess. Based on peer review of innovation projects, SBIR agencies distribute funds to typically five to seven times more early-start projects than venture capital, and as a result SBIR devotes far more manpower to the screening of projects than any individual venture capitalist can afford. For this reason, the awarding of an SBIR grant itself becomes an informative "signal" to venture investors about the otherwise hard-to-gauge quality of the project, and SBIR awardees are more likely to receive venture capital funding than non-awardees. In the life sciences, for example, Keller and Block found that about 20 percent of venture capital investment goes to firms that have been SBIR awardees, even though SBIR funding only amounts to 2.5 percent of total federal R&D funding. Besides its reviewing manpower, SBIR further bridges the insider information gap for venture capitalists by organizing various workshops and conferences that allow firms, investors, and government contractors to establish deeper networks of communication.

In the same vein, Feldman and Francis' (2003) study of the biotechnology cluster in the US Capitol region also highlighted the indispensible role of the state in smoothing the process of innovation financing. Besides Silicon Valley in California and Route 128 in Massachusetts, the US Capitol region covering portions of Maryland, Virginia, and the District of Columbia has emerged as the region with the third largest concentration of bioscience companies in the US. Bypassing the traditional route of private equity, public funding (often in the form of procurement contracts from federal and state agencies) has played a key role in financing biotech innovation in the Capitol region. Maryland, for example, ranked fourth out of all states in the number of the above-mentioned SBIR awards between 1995 and 2000, and second (only to California) in the amount of federal R&D obligations, and third in the amount of federal funding received, through departments such as Defense, Agriculture, Commerce, Health and Human Resources, and National Aeronautics and Space Administration. The R&D budget from a single public institution (the National Institute of Health, located in Bethesda, Maryland) alone, for example, was $23.56 billion in 2002, which was *330 times* the total amount of venture capital raised for Maryland's biotech sector ($69.6 million) (Feldman and Francis 2003, pp. 769–70). These case studies of innovation financing in the US suggest that the real contribution of venture capital to high-tech startups in the US may be smaller than can be inferred from the sheer size of the venture capital market.

Just as the role of venture capital may be less important than commonly expected in liberal market economies, various case studies have also highlighted the surprising success of venture capital in stimulating innovation in strategically coordinated economies. Deeg's (2009) comparison of equity financing across European countries, for example, identified a sharp contrast between the strong emphasis on public equity (IPOs) in the UK (a liberal market economy) on one hand, and the strong emphasis on private equity (venture capital and buyouts) in Germany, France, and Italy (all coordinated economies) on the other. In a more focused comparison of two biomedical regions (Munich in Germany and Cambridge in the UK), Crouch et al. (2009) found that, although Germany's own traditional bank-based model proved, as expected, to be ill-adapted to the financing of risky high-tech innovation, Munich's biotech firms were able to overcome this hurdle by securing funding not only from the state but also, more important to the book's theme, international venture capital markets. As a result, Munich has outperformed Cambridge as a center of biomedical innovation. The appeal of German firms to international venture capitalists not only underlines the quality of insider information that the Germany economy can offer to investors, but also goes to some extent in resolving a puzzle noted earlier in this chapter: how could venture capital prove so effective in promoting innovation in European countries, when their own venture capital markets are all so small? The answer, it appears, is that as long as the economy can provide high-quality insider information, it can attract venture capitalists from far beyond its own borders.

Besides providing concrete examples of how nations differ in the performance of their financial markets, evidence from the case-study literature also sheds light on the causal mechanism underlining this outcome: liberal market economies enhance public information; strategically coordinated economies enhance insider information. Returning to the example of German biomedical innovation, Krauss and Wolf's (2002) examination of the Rhine-Neckar-Triangle region demonstrated that the successful financing of biotech firms in this region was closely linked to BioScience Venture Company, a seed capital fund. This fund brings associate members not only from the major companies but also from Sparkassen/banking institutions, all of which represent key nodes in the region's traditional patient capital network and help generate valuable insider information for the venture capital fund. A similar integration of bank and venture capital financing to draw upon insider information is also at play in Germany's television film-making industry centered on Cologne (Crouch et al. 2009). Contrary to the common association of banks with low-risk projects, local saving banks in the Cologne region are willing to take on high-risk projects that are typical of the media industry, because of the deep links they have with the companies. In other words, by financing *high-risk* projects on the basis of deep *insider information*, saving banks in the Cologne region become what Crouch et al. (p. 669) refer to as "a sort of venture capital." This logic of coordination and insider information is also reflected in the bank-based financing of a more incrementally innovative sector: the furniture industry. As Rafiqui (2010) found for Sweden, the success of this industry in the country's Tibro region depends on a highly reciprocal financial network centered on bank financing. The highly localized system of production, within which bank loans and bills of exchange are circulated, lowers communication and monitoring costs, and increases the quality of insider information that banks have access to. I will return to Tibro's furniture industry later in the book, because it also provides important case evidence regarding why strategically coordinated economies tend to prefer process innovation to product innovation, an important theme in Chapter 3.

Just as dense firm–investor coordination is associated with *insider* information, the growth of *public* information often goes hand in hand with the decline of such coordination. Such a process of transition to public information is clearly illustrated in Frédéric Widmer's (2011) description of the machine industry in Switzerland. In financial transparency, the country's national standard on corporate public disclosure is much less stringent than international standards, in effect favoring investors who consume insider information (lenders) at the expense of those who depend on public information (public equity investors). Since the early 1990s, however, Swiss firms have gradually adopted the more stringent International Financial Report Standard (IFRS) or the United States Generally Accepted Accounting Principles (US-GAAP), due to international pressure for standardization, exerted in particular by the New York Stock Exchange, on which many Swiss firms are listed. For example, all of the nine largest Swiss machine firms examined by Widmer had adopted either IFRS or US-GAAP by 2001. However, while international accounting standardization helped increase the quality of public information on the Swiss financial market, it also led to the destruction of the interlocking bank–industry nexus which traditionally underpinned the Swiss machine industry. As Widmer pointed out, the participation of bankers on the boards of machine firms used to be the crucial mechanism through which investors gain privileged access to "opaque information" (p. 684), but as greater transparency in public information crowds out such insider information, the utility of such bank–industry links also declined. In 1995, there were twenty-six directors across machine firms that established up to thirty-three industry–bank links with the three main banks (Crédit Suisse, Société de Banque Suisse, and Union Bank of Switzerland), but by 2005 twelve of them had left their firms and thirteen left their banks.

Finally, Gregory Jackson's (2009) study of corporate governance in Japan is also a valuable window into the causal processes at work in this chapter. As noted earlier in my quantitative analysis, scholars have long pointed out that the real varieties of capitalism are far less stylized than a simple distinction between coordinated and liberal market economies. In reality, both of these two broad categories have many subcategories of their own. Jackson's comparison of three subcategories *within Japan* is an example of such *subnational* variation which cannot be captured in my quantitative analysis, which relies on national-level indicators of coordination. The first subcategory, or the J-type, represents the traditional mode of *relational* finance and *long-term* employment, most commonly found in sectors such as construction, chemicals, machinery and automobiles (small firms), and textiles. The second subcategory, or Type I hybrid, combines *capital-market-based* finance and *long-term* employment, which represents many well-known firms in Japan such as Toyota, Canon, Kao, Hitachi, DoCoMo, and some firms in the Mitsubishi group. The third subcategory, or Type II hybrid, has the opposite combination of *relational* finance and *short-term* employment, in sectors ranging from retail to IT services. In other words, firm–investor relationship in Type I hybrid is more liberal market-based than both the traditional J-type and Type II hybrid. As Jackson pointed out, the quality of public information provided to the financial market is also notably higher for Type I hybrid firms than others, especially in terms of the efforts firms make to promote corporate transparency and disclosure. Such distinction can also be found, in an even more nuanced way, within the Type I hybrid group, between "blue-chip" firms that depend relatively more on market-based financing (such as Toyota) and those that rely more on relational financing (such as DoCoMo). Compared with the blue-chips, companies in the latter category are also more cautious about promoting corporate transparency and public disclosure. In other words, the stronger the firm–investor ties, the more important is insider information relative to public information.

Conclusion

In this chapter, I have argued that different types of capitalism specialize in communicating different types of information. *Strategically coordinated capitalism communicates insider information but suppresses public information, and vice versa for liberal market capitalism.* Given this contrast between liberal and coordinated capitalism, whose investors are better hunters of innovation (as this chapter's title asks)? In markets operating on public information (such as the stock market), liberal market economies produce better hunters of innovation; in markets relying on insider information (banking and venture capital), coordinated economies create better hunters instead. In other words, "how firms speak" in financial markets has genuine consequences for the ability of investors to perform effectively. By encouraging firms to send public rather than insider information, liberal market capitalism creates a *minority protection effect* on the financial market: it shifts the informational advantage from major investors (such as banks and wealthy venture capitalists) to minority shareholders (the millions of individuals and households investing in the stock market).

Besides the book's overarching theme of information and capitalism, this chapter's findings also have some implications for another commonly discussed theme in the comparison of capitalism: competition (Katzenstein 1985; Kogut 1993; Best 1990, 2001; Lazonick 1990). Starting from Peter Katzenstein's (1985) classic work on how small open economies adjust to international competition, the neocorporatism and varieties of capitalism literatures have highlighted how deep and long-term cooperation may facilitate rather than suppress effective competition. In particular, firms in coordinated capitalism not only compete strongly on the international stage but also pick "good competition" ("race to the top" in product quality) at the expense of "bad competition" ("race to the bottom" in costs and wages) (Hall and Soskice 2001; Hollingsworth and Boyer 1997; Streeck and Thelen 2005). This chapter adds a new angle to this debate about bad competition, in the form of "rat races." While races to the bottom may have bad consequences (poorer training, lower wages, and more poverty), they also have benefits, pushing the frontier in cost and price competitiveness. Rat races, by contrast, do not even move the frontier forward: when financial markets communicate poor information, investors may have to keep running (on the amount of investment offer), just to stand still (on the quality of innovation projects). For this reason, compared with races to the bottom, rat races may be an even more perverse form of bad competition. Seen in this light, the chapter's findings imply that, in reality, neither strategically coordinated nor liberal market economies can avoid bad competition: rat races hit stock investors in the former and venture capitalists in the latter. It is important to note that these findings will have various interesting implications for public policy. In order to provide a common framework for a concentrated analysis, I use the book's conclusion chapter to discuss all public policy implications from the book.

This chapter has examined the distribution of information through a framework of "how firms speak" (to investors in financial markets). Besides firms' hidden financial information, valuable information also exists in many other forms. For example, when firms succeed in R&D, they create new information, in the form of new technological knowledge. Should they use this knowledge to bring new products to the market, or develop new ways of raising productivity in the production process? This choice between "product innovation" and "process innovation," and its implication for employment, is the key topic of the next chapter.

3

Whose Innovation Creates More Jobs?

Out of the many yardsticks for economic performance, few touch our daily lives as concretely as jobs. For this reason, topics such as unemployment, job creation, and productivity are not only central themes in academic and policy discussions, but also perpetual battlegrounds in electoral politics and ideological debates. Although the comparative political economy literature has devoted a vast body of work to the topic of employment (Scharpf and Schmidt 2000; Regini and Esping-Andersen 2000; Zeitlin and Trubek 2003; Pontusson 2005; Mares 2006; Bradley and Stephens 2007; Clasen and Clegg 2011; Iversen and Rosenbluth 2010), the impact of technological innovation on employment has not been widely examined in this literature.

Does technological innovation create or destroy jobs? The obvious answer is that it does both. New advances in technology are one of the most important examples of "creative destruction" (also known commonly as "Schumpeter's gale") that scholars since Karl Marx (1867) have associated with wealth accumulation in capitalist economies (Schumpeter 1942; Kondratieff 1984; Castells 1996; Harvey 2010). Each wave of new technologies creates its own Luddites who lose markets and jobs to newcomers (Mokyr 2002).

Although innovation may create as well as destroy, this chapter shows that innovation leads to different outcomes in different varieties of capitalism. While innovation in liberal market economies is job-friendly, innovation in strategically coordinated economies is job-destructive. The logic behind this pattern lies in the different ways firms utilize their innovation output in different types of capitalism. New technological knowledge from innovation may lead to either new products ("product innovation"), or new methods of increasing productivity ("process innovation"). Product and process innovation generate different types of returns for the innovator. In the former, new products create a temporary monopoly, allowing the innovator to earn a premium for novelty, in the form of "monopoly rents." In the latter, new technical knowledge raises productivity (smaller overhead, quicker turnover, and higher output), allowing the innovator to capture "productivity rents" during the production process. The premium for novelty or monopoly disappears when the innovation is shared with other firms. The ability to lift productivity, by contrast, does not decline when firms share process innovation with each other. In fact, due to increasing returns and network complementarities, sharing process innovation may further increase productivity for all involved. In other words, while mutual sharing of product innovation among firms may *eliminate* their monopoly rents and make *no one* better off, the relational sharing of process innovation may *multiply* their productivity rents and make *everyone* better off. As a result, depending on relational depth, different varieties of capitalisms will specialize in different types of innovation. Firms in strategically coordinated capitalism will prefer process innovation because it rewards sharing with relational insiders; firms in liberal market capitalism will prefer product innovation because it is more valuable when firms do not commit to relational sharing.

Because product innovation introduces new lines of products, it opens up new markets and creates new job opportunities; by contrast, because process innovation increases productivity and allows more products to be made faster with less input, it is labor-saving and detrimental to employment. As a result, product innovation in Anglo-Saxon countries brings new job opportunities, while process innovation in Europe eliminates jobs. In other words, although innovation is creative destruction, creation and destruction balance out differently across different varieties of capitalism. The more strategically coordinated capitalism, the more innovation in "process" rather than "product," and the more it destroys rather than creates jobs.

It is important to point out the limit to the scope of this argument. Because this is a theory about product and process innovation, it relates patterns of employment specifically to the introduction of new products on the market, and new productivity-raising techniques on the production floor. As a result, this theory is more likely to hold true for goods-producing, especially manufacturing, sectors than nonindustrial service sectors such as retail, childcare, and social services, where there is less room for rapidly increasing productivity or developing new products. Instead, employment in these service sectors is more directly determined by how fast they can adapt to declining demand for low-skill labor (a common consequence of skill-biased technological change (Berman et al. 1998; Autor et al. 1998), either on the downside (lessening downward wage rigidity, employment protection, and work-pattern regulations) or the upside (intensifying social investment in human capital, such as higher-quality public education and labor market activation). In reality, high unemployment hits some European economies (Christian democratic countries) much harder than others (Scandinavian countries), because these two sets of countries differ notably in the nature of their public and private service sectors as well as labor market and welfare state institutions. In service sectors, as Thelen (2014) points out, there are significant differences between the rigid insider/outsider "dualism" (Palier and Thelen 2010) in Christian democratic countries and the flexible inclusion of outsiders in Scandinavian countries. By more fully integrating the various vulnerable margins of the labor force, Scandinavian countries manage to achieve far greater success in job creation than their Christian democratic counterparts. Because product and process innovation are more relevant for manufacturing than services, the chapter's findings cannot truly reflect the various rich insights about employment in public and private service sectors. Although my findings on employment may be more relevant to manufacturing than services, this does not necessarily imply that the findings are less important. Manufacturing continues to be a key component of export industries, and for that reason, a key determinant of a country's competitive position on the international market. Therefore, the impact of innovation on jobs will have direct implications for policy-makers across advanced industrialized countries, as they seek to balance the twin goals of more jobs versus more industrial competitiveness. For continental European countries, in particular, the finding that their technological innovation *destroys* jobs within the *innovating* sector exacerbates the job-creation problem they already face. Since these economies already experience great difficulties in creating low-end jobs in the service sector (Iversen and Wren 1998), continued generation of high-end manufacturing jobs, through ongoing technological innovation, could have been an important alternative path to jobs. This chapter's findings imply, however, that innovation may in fact further aggravate continental Europe's unemployment problem.

More jobs alone are insufficient for sustaining economic growth in the long run. Both human and physical capital inputs are subject to diminishing returns. Unless the economy can offset diminishing returns by continuously raising productivity through innovation, more input can only extract ever less economic growth over time (Solow 1957; Jorgenson 1995; Rosenberg 1996; Lau 1996; Griliches 2000). Therefore, the economy not only needs employment-friendly innovation to fuel the labor market, it also needs productivity-friendly innovation to fuel its own growth in the long run. However, the more productive the technology, the less input the production requires, so there is an inherent tradeoff between employment and productivity as two goals of innovation. This chapter provides clear evidence for this tradeoff across affluent capitalist economies. While Anglo-Saxon countries compare favorably against their European counterparts in the job effect of innovation, the records are stacked the other way in productivity. Process innovation in coordinated European economies may have curtailed their ability to provide jobs, but by enabling more value to be created from less input, innovation has also allowed them to more rapidly lift productivity in the economy.

In short, innovation in strategically coordinated economies destroys jobs but enhances productivity; in liberal market economies, by contrast, new technologies increase employment but are less capable of lifting productivity. I refer to the job-expanding effect of innovation in liberal market capitalism as the *extensive-growth effect*: in these countries, innovation pushes the economy on the *extensive* margin, expanding product lines, creating new markets, and generating new job opportunities. Correspondingly, I refer to the productivity-deepening effect of innovation in strategically coordinated capitalism as the *intensive-growth effect*: in these countries, innovation pushes the economy on the *intensive* margin, deepening the productivity and sophistication in the making of existing products.

This chapter has two main sections. In the first section, I introduce the distinction between product and process innovation and explain how they create different sources of rents. On this basis, I connect the different rents to different relational depths in coordinated and liberal market economies. I provide empirical evidence that the more coordinated the economy, the more firms engage in process rather than product innovation. I draw out the full implications for employment and productivity in the second section. Based on pooled time series analysis of innovation and employment data, I show that innovation is more job-friendly in liberal market capitalism. Then, I show that records are stacked the other way on productivity: innovation is more effective in raising productivity in strategically coordinated capitalism.

Product and Process Innovation

Beyond Radical and Incremental Innovation

One of the most influential frameworks for classifying technological innovation is the distinction between radical and incremental innovation. Various theoretical and empirical works have referred to this distinction in their comparison of affluent capitalist economies (Ornston 2012; Taylor 2004; Akkermans et al. 2009; Herrmann 2008; Lange 2009; Casper 2009). Although the radical/incremental framework has received dominant attention in the varieties of capitalism literature, it is but one way of cataloguing innovation. To shed light on why different capitalisms may differ in the employment impact of innovation, I introduce another typology, based on whether the innovation is in the *production process* or the *actual product*. Although the distinction between "product innovation" and "process innovation" occupies an important place in the general literature on technological innovation (Utterback 1994; Evangelista 1999; Edquist, Hommen, and McKelvey 2001; Powell and Grodal 2005), it has not been widely explored in the comparative political economy literature. I start by explaining the conceptual difference between product innovation and process innovation. This paves the ground for a survey-based measure of product and process innovation later in the empirical analysis.

Product innovation leads to new products, while process innovation leads to new methods of raising productivity in the production process (Edquist et al. 2001). Based on this conceptual distinction, it is clear that the product/process divide cuts across, rather than overlaps with, the radical/incremental divide in classifying innovation. A product innovation may lead to either considerably different products (such as Mac OS X vis-à-vis Windows operating systems) or incrementally better products (such as "mild hybrid" vis-à-vis "full hybrid" car engines). Similarly, a process innovation may be relatively drastic (such as the invention of the "hot blast" technique in iron production and the "float glass" technique in glass production, both of which led to revolutionary cost efficiency improvements for their respective industries (Landes 1969; Mokyr 1990)) or gradual (such as the growing use of platform sharing in automobile engineering and design (Gnosh and Morita 2008)).

Of course, in reality, product and process innovation rarely occur in the absence of each other. On the contrary, they are complementary. Process innovation breeds product innovation: a new production technique may not only raise productivity in the production process but also pave the way for the creation of new or better products. Similarly, product innovation also breeds process innovation: the development of a new product often opens up new horizons in the search for methods to improve the production process (Evangelista 1999; Bartel et al. 2007). Because of the inherent complementarity between product and process innovation, it may be unlikely to observe them as "tradeoffs" against each other. However, this is precisely what the chapter will show: countries that do more product innovation tend to do less process innovation, and vice versa. Although from a purely technological perspective these two types of innovation should reinforce each other, one of them *penalizes* sharing among firms, while the other has the opposite effect of *rewarding* such relational commitment. As a result, capitalisms with different coordination capacity will choose differently between these two otherwise complementary activities. In other words, the inherent technological affinities between product and process innovation should pose a relatively hard test for the book's core argument that differences in coordination depth drive capitalisms apart in their patterns of innovation. What brings product and process innovation into conflict is the difference in the nature of rents they generate.

Monopoly and Productivity Rents

All successful innovations result in some form of new technological knowledge. However, product and process innovation differ in the way they allow the innovating firm to profit from its own new technological knowledge. Product innovation, by definition, leads to new (either drastically or incrementally) products on the market. By being new, the products allow the innovator to create a temporary monopoly, and correspondingly set prices above competitive cost (Chaney et al. 1991; Doukas and Switzer 1992). For example, iPhones were able to price above equilibrium market valuation because Apple Inc. did not immediately face serious competitors offering comparable lines of products (Korkeamäki and Takalo 2012; Chen et al. 2005). This price premium, it is important to note, is not necessarily unbounded. For example, as Baake and Boom (2001) point out, when consumers regard the entire network size of a new product as an element of its quality (i.e. how useful a product is depends on how many people use it), all firms will seek to occupy a larger slice of the network. As "competition for market share" becomes more important relative to "raising the price mark-up," there will be a limit to how high firms can set their novelty premium. More generally, the price premium on novelty may be either large (if the innovation is radical) or moderate (if the innovation is incremental). Regardless of its size, the novelty premium flows from "being new": consumers are willing to pay a higher price because they can't find it anywhere else in the market. In other words, if the product innovation is shared with other firms, novelty disappears, and the innovator loses her temporary monopoly rents (Galbraith and Berner 2001). This implies that product innovation creates an indivisible prize: if the output from product innovation is shared with other firms, its novelty rents are dissipated, and as a result *no firm* is able to collect these innovation rents.

On the other hand, process innovation generates a different type of rents. Because process innovation leads to new methods of raising productivity in the production process, the innovator can push marginal costs below competitive prices, through smaller overhead, quicker turnover, and more flexible supply-chain coordination, among others (Edquist et al. 2001). Again, the profits from higher productivity may be either large or moderate, depending on whether the innovation is radical or incremental, but, regardless, process innovation creates rents from *cost efficiencies in the production process*, in contrast to product innovation, which creates rents from *monopoly over the new product offering*. In other words, while the "novelty rents" from product innovation are reaped only after the new product is brought to the market, the "productivity rents" from process innovation are realized even before the product reaches the market. As a result, while the loss of product monopoly will eliminate the rents from product innovation, it does not prevent the continuing gains in cost efficiency from process innovation.

For this reason, unlike rents from product innovation, productivity rents from process innovation are not eliminated when the new technological knowledge is shared with other firms. One firm's ability to reduce overheads, increase production speed, and streamline inventory supplies using a process innovation does not reduce another firm's ability to use the same process innovation to achieve the same kind of cost efficiency. In fact, instead of dissipating productivity rents, the pooling of process innovation among firms may further increase productivity rents for all involved. Because R&D often benefits from network complementarities, platform sharing, and increasing returns, the total productivity return from pooling individual process innovations may be greater than their sum (Graf 2006; Powell and Grodal 2005; Saxenian 1994). As a result, a relational commitment to the sharing of innovation output may lead to very different outcomes, depending on whether it is product or process innovation. While the sharing of product innovation among firms *eliminates* every participant's monopoly rents and makes *no one* better off, the relational sharing of process innovation *multiplies* every participant's productivity returns and makes *everyone* involved better off. As a result, depending on relational depth, different varieties of capitalisms will specialize in different types of innovation. Firms in strategically coordinated capitalism will prefer process innovation because it rewards sharing with relational insiders; firms in liberal market capitalism will prefer product innovation because it is more valuable precisely when firms cannot commit to relational sharing.

> H1: *the more coordinated the economy, the more firms engage in process relative to product innovation.*

Next, I take hypothesis H1 to test, and demonstrate that countries do indeed choose differently between product and process innovation, depending on the coordination depth of their economy. This paves the ground for the second section of the chapter, where I draw out the full implications for employment and productivity.

Empirical Analysis

The testing of hypothesis H1 in the varieties of capitalism framework requires data that is not only comparable across a wide range of OECD countries but also capable of transparently separating product innovation from process innovation among firms' innovation activities. For such cross-national and detailed data on firms' innovation strategies, I turn to the Community Innovation Survey (CIS), jointly administered by Eurostat and the Directorate-General Enterprise of the EU. The firm survey covers most of the EU states plus Norway, and includes all enterprises with ten or more employees. Starting from its third wave in 2001, this nonpanel survey has followed a standardized framework for its questionnaire, based on the *Oslo Manual* (OECD 2005), the internationally recognized methodological guide for measuring innovation. In order to preserve the anonymity of individual enterprises surveyed, the data is released to the public on the sector (NACE Rev. 1/Rev. 2) level rather than the firm level. In the absence of direct observation about individual firms on a micro level, the researcher has to construct measures that are sector-level correlates of firm choices between product and process innovation.

Starting from its third wave in 2001, the survey began to pose standardized questions that explicitly ask firms to report the extent to which their technological innovation leads to improvements in products, or alternatively, improvements in the production process. In total, there are four waves (the third through the sixth wave, respectively in 2001, 2004, 2006, and 2008) that asked sufficient questions to cleanly distinguish product from process innovation. However, because these are not panel surveys, and because there are very few years for observation (four survey waves) relative to three times as many cross-sectional units (up to twelve countries),[1] I analyze data from each survey as a separate cross-sectional analysis and use the findings from different surveys as robustness checks against each other, rather than pooling all four surveys to run a single time series analysis.

In the 2001 survey, one question asks firms to rate by how much innovation has increased the range of products on offer (choosing from three categories "low," "medium," and "high"). Individual enterprises' responses to these questions are not available, but CIS does provide, for each sector, the percentage of firms choosing each of the three categories for their answer. For each sector, I score the answer's categories as respectively "one" for "low" increase in product range, "two" for "medium," and "three" for "high." Then, I weight each score with the percentage of firms choosing that score. Finally, I sum the weighted scores to derive an indicator for the sector's overall extent of product range improvements due to innovation. Another question in the survey asks, in similar wording, about the extent of increase in product quality due to innovation. I implement the same procedure: first score the answer's categories into one through three, then weight the scores by the percentage of firms choosing them, and finally sum the weighted scores to create an indicator for the overall extent of product quality improvements due to innovation. Because both questions ask firms to evaluate the extent to which innovation leads to improvements in products, both serve to measure product innovation. Since a strong Cronbach's α (0.71) suggests that these two indicators do indeed tap the same underlying concept, I average them to produce an overall, sector-level, index for the intensity of product innovation.

---

[1] These countries are Austria, Belgium, Denmark, Finland, France, Germany, Ireland, Italy, the Netherlands, Norway, Sweden, and the UK.

I use the same scoring and weighting procedure to derive a sector-level index for the intensity of process innovation. This index is again based on two source indicators, each tapping a specific type of improvement to the production process due to innovation, using again the three answer categories ("low," "medium," and "high"). One asks firms to report the extent of increase in the flexibility of the production process through innovation, while the other asks about the extent of reduction in production cost through innovation. Again, the two indicators extract information from the same latent concept (Cronbach's $\alpha=0.88$), so I average them to gauge the overall intensity of process innovation in each sector. Hypothesis H1 asks not so much about the intensity of product or process innovation per se, but instead about how they compare relative to each other. To reflect this relative comparison, I use the ratio of the intensity for process innovation over the intensity for product innovation as the dependent variable for H1. I also follow a similar procedure in constructing the dependent variable from the 2004, 2006, and 2008 surveys.

Consistent with the previous chapter, I continue to use the Hicks–Kenworthy index to measure coordination, and check the robustness of findings with the Hall–Gingerich index as an alternative measure of coordination. Besides coordination as the core independent variable, I also include several control variables that may further affect firms' choices between product and process innovation. Because the theories underlying these control variables are not related to this chapter's main arguments, I only briefly outline their logic here, and refer readers to the relevant literature that discusses these issues in greater detail. Technical knowledge used during R&D can be either "embodied" in fixed investment (new machinery or equipment), or "disembodied" in its raw form as pure information, passed on through R&D personnel, publications, or patent disclosure (Dollar and Wolff 1993; Pianta 1998; Spiezia and Vivarelli 2002). There are many factors that may affect whether firms choose to invest in embodied or disembodied knowledge, such as compatibility with existing stock of fixed assets, conditions in the labor market, and laws governing patents and publications. While disembodied knowledge is key to propelling product innovation, embodied knowledge is an important contributor to process innovation. The literature points out several reasons for this (Evangelista 1999). For instance, disembodied abstract knowledge is more important than fixed investment in the early stage of the product life cycle, where product innovation tends to dominate. On the other hand, fixed assets that embody technical knowledge are not inventoried as end products; instead, they are plugged into production as intermediate inputs to increase the productivity of the production process. CIS has two variables that measure enterprise expenditure on, respectively, the acquisition of external knowledge and the purchase of new machinery/equipment/software, both calculated as a share of the sector's total turnover. I use the ratio of expenditure on external knowledge acquisition over expenditure on machinery/equipment/software purchase to capture the extent to which firms invest in disembodied relative to embodied knowledge, which may affect their capacity to engage in product relative to process innovation.

Besides the "medium" for its diffusion (embodied or disembodied), the "source" of technical knowledge used in R&D may also affect whether it is suitable for product or process innovation, depending on whether the knowledge source is on or off the actual site of production. Where information used in R&D is acquired from sectors far removed from the production process, such as educational institutions or government laboratories, the general and abstract nature of the knowledge may limit the extent to which it can be applied directly to concrete production processes, and therefore limit its effect in raising productivity. However, precisely because it is less bound by the concrete requirements of existing production, such open-ended knowledge may also help firms widen their horizons for the imagination of new products and new markets (Nelson 1996; Lawton-Smith 2006). Firms may come into contact with such off-site technological information for many reasons. For example, when firms face capital constraint as a result of limited external financing, they may cut setup cost by contracting R&D out to universities or other higher-education institutions. Similarly, firms located in industries where the government plays a heavy role (such as transportation and health) may benefit directly from absorbing parallel research conducted by the state. As firms draw more heavily from state- or university-supplied research for innovation, they may become more effective in creating new products than improving the productivity of existing production processes, and correspondingly find product innovation more attractive than process innovation. CIS has two variables that measure each sector's percentage of firms reporting intensive use of information from, respectively, "universities and other higher education institutions" and "government or public research institutes"[2] during innovation. I use the sum of these two variables to measure the extent to which firms use off-site information for innovation.

Sometimes, a downstream firm subcontracts the R&D task to another firm upstream. The *Oslo Manual* defines such innovation as "extramural." Regardless of whether the innovation leads to new products or production processes at the end of the vertical chain, the upstream firm faces a product innovation, because its innovation output is sold directly as a product to the downstream firm. Therefore, the more firms spend on extramural R&D to subcontract innovation to other firms, the more innovation is carried out as product rather than process innovation. To incorporate this factor, I adopt the CIS variable

---

[2] In the 2001 survey, this category is worded slightly differently, as "government or private nonprofit research institutes."

that measures overall firm expenditure on extramural R&D, calculated as a share of sector total turnover. Furthermore, depending on the extent to which end-users, downstream firms, and their upstream suppliers share the same technological competence and operate in the same market, some firms focus more on communicating downstream with clients during innovation, while others focus more on communicating upstream with their suppliers. As the "systems of innovation" literature (Nelson 1993; Lundvall 2010; Edquist 1997; Edquist and McKelvey 2000) argues, while downstream client–producer communication is particularly helpful for the development of new products in the early stage of the product cycle, upstream producer–supplier communication is particularly effective in stimulating process innovation in the later stage of the product cycle (Andersen and Lundvall 1988). CIS has two variables that measure the percentage of firms in each sector citing respectively "suppliers of equipment, materials, components or software" and "clients or customers" as an intensive source of information for their innovation. I use the ratio of percentage citing "customers" over the percentage citing "suppliers" to measure the extent to which firms rely on downstream relative to upstream communication. The busier the downstream channel of communication, the more firms are likely to engage in product rather than process innovation.

Finally, firms may also adopt purely social and organizational changes that enhance the production process without generating substantially new technological knowledge (such as Quality Circle as a flexible and friendly environment for communicating Japanese workers' individual suggestions for product improvement) (Kenney and Florida 1993). Although the new organizational strategies are themselves non-technological, they can act as a catalyst for parallel development in technologies that serve the same purpose of improving the production process (Lazonick 2005). Since organizational innovations may reinforce process innovations, I include the CIS variable that measures the percentage of firms in each sector that undertake new organizational strategies.

I use Seemingly Unrelated Regression (SUR) (Zellner 1962) to estimate a pair of equations, respectively with and without sector effects. The twin equations are estimated jointly because they share a common dataset and all control variables, with errors correlated across them. This procedure allows for simultaneous estimation of multiple equations with (nearly) identical variables, a common set of data, and, for this reason, interdependent errors between equations. This SUR technique of analyzing survey-based innovation data follows other works in the innovation literature, such as Betts (1997) and Piva et al. (2005).

As noted earlier, in order to make the findings more robust, I analyze the survey data using alternatively the Hicks–Kenworthy and Hall–Gingerich measure of coordination. With two equations for each SUR estimation, four surveys, and two different measures of coordination, this exercise produces in total sixteen sets of findings that may serve as robustness checks against each other. In order to ensure concise presentation while still providing a flavor of the findings, the eight columns in Table 3.1 alternate between the two measures of coordination, using Hicks–Kenworthy for even-number ordered surveys (third and fifth waves) and Hall–Gingerich for odd-number ordered surveys (fourth and sixth waves).

I start the discussion with the 2001 survey (first two columns). Across countries in this survey, the average ratio of process innovation over product innovation is 1.13. As the coefficients for coordination indicate, a one-unit increase in coordination will raise this ratio by around 0.9. In other words, this expansion in the relative proportion of process innovation is equal to as much as 80 percent of the average extent to which firms engage in process relative to product innovation across the countries. This, of course, is a very sizeable effect, but it should be put in a proper prospective. A one-unit movement along the Hicks–Kenworthy index will encompass its entire range (from zero to one), so the raw coefficients capture an extreme scenario of moving directly from "minimum" to "maximum" coordination. In reality, the difference between most countries in the strength of coordination is much less drastic, and for this reason it is important to interpret the findings on the basis of more realistically scaled comparisons.

Each standard deviation of increase in coordination, for example, tilts the innovation mix from product to process by 0.30, or about a third of the average extent to which countries engage in process relative to product innovation. Since the mean Hicks–Kenworthy coordination score across the countries is 0.51, this implies that an economy with average coordination will engage in more process innovation than an economy with minimum coordination by an amount that is almost equal to half of the average proportion of process innovation across countries.

With all control variables at their mean, a country with median coordination will mix process and product innovations on the basis of roughly one-and-a-half to one (1.56 on the ratio). A standard deviation more coordination will increase the ratio to 1.86, while two standard deviations above will result in more than twice as much (2.16) process innovation than product innovation. Moving in the other direction, by contrast, two standard deviations below average coordination will result in an almost even split (0.96 on the ratio) between the two types of innovation.

To gain some further substantive meaning from the results, I use the point estimates to calculate the difference in the product/process innovation mix across different countries, ceteris paribus. For example, if British firms (0.10 on the Hicks–Kenworthy index) are endowed with coordination institutions of Austrian strength (0.96 on the index), they will add another 0.81 to the ratio

Table 3.1. Intensity of Process Relative to Product Innovation, Twelve OECD Countries

Community Innovation Survey Wave	Third Wave (2001) Equation a	Third Wave (2001) Equation b	Fourth Wave (2004) Equation a	Fourth Wave (2004) Equation b	Fifth Wave (2006) Equation a	Fifth Wave (2006) Equation b	Sixth Wave (2008) Equation a	Sixth Wave (2008) Equation b
Coordination	0.95(0.18)***	0.91(0.19)***	1.49(0.44)***	1.32(0.45)***	0.49(0.15)***	0.53(0.11)**	0.85(0.25)***	0.78(0.16)***
Extramural R&D	−0.41(0.14)**	−0.38(0.13)***	−0.55(0.13)***	−0.59(0.20)**	−0.60(0.24)***	−0.51(0.22)**	−0.72(0.21)**	−0.75(0.20)***
Universities/government as source of information	−0.06(0.03)**	−0.08(0.04)*	−0.1(0.02)***	−0.07(0.03)*	−0.06(0.02)***	−0.04(0.02)	−0.13(0.04)**	−0.09(0.00)***
Organizational innovation	0.39(0.13)**	0.46(0.11)***	0.22(0.08)***	0.28(0.10)***	0.54(0.12)***	0.48(0.21)*	0.50(0.20)**	0.61(0.32)
Investment in disembodied relative to embodied knowledge	−0.40(0.32)	−0.46(0.19)**	−0.62(0.29)*	−0.72(0.35)*	−0.39(0.30)	−0.35(0.14)**	−0.48(0.19)**	−0.45(0.21)*
Intensity of downstream relative to upstream communication	−0.12(0.44)	−0.18(0.47)	−0.34(0.26)	−0.51(0.50)	−0.31(0.21)	−0.36(0.50)	−0.55(0.40)	−0.79(0.91)
Constant	1.18(0.46)***	1.16 (0.39)***	1.26(0.18)***	1.2(0.13)***	1.19(0.25)***	1.56(0.20)***	1.48(0.16)***	1.40(0.32)***
N	64	64	150	150	71	71	194	194
Parms	15	12	32	12	40	12	62	12
RMSE	0.18	0.19	0.18	0.19	0.16	0.16	0.18	0.18
P	0.00	0.00	0.00	0.00	0.00	0.00	0.00	0.00
R-squared	0.48	0.42	0.46	0.39	0.45	0.43	0.39	0.38

* $p<0.1$; ** $p<0.05$; *** $p<0.01$

of process over product innovation, equivalent to about 70 percent of the average extent to which countries choose process over product innovation. On the other hand, if coordination in Germany (0.80 on the Hicks–Kenworthy index) weakens to the level of its French counterpart (0.40 on the index), firms will switch instead to more product innovation, reducing the ratio of process over product innovation by 0.38, or about 33 percent of the average weight countries put on process innovation. Similarly, Danish (0.72 on coordination) firms' focus on process relative to product innovation is stronger than their Italian (0.44 on coordination) counterparts by about a third of the cross-country average. On the other hand, technological innovation in weakly coordinated Ireland (0.07 on coordination) is more oriented towards new products than strongly coordinated Sweden (0.97 on coordination) by more than 60 percent of the cross-country average. In short, as hypothesis H1 suggests, the more coordinated the economy, the more firms choose process over product innovation.

This core pattern of finding persists through the other surveys (2004, 2006, and 2008) reported in Table 3.1, which alternate between Hicks–Kenworthy and Hall–Gingerich as the measure of coordination. Looking across all four surveys, coordination has a consistently positive and statistically significant effect on the relative intensity of process over product innovation. Depending on the year of the survey, one unit of increase in coordination on the Hicks–Kenworthy or Hall–Gingerich index will add somewhere between 0.49 and 1.5 to the ratio of process over product innovation in the economy. A more realistically scaled one standard deviation of change in coordination can still translate into a difference of up to 0.5 on the ratio, which is sufficient to transform an even split (i.e. ratio=1) between process and product innovation into a mix of one-and-a-half to one (i.e. ratio=1.5), or alternatively, make process innovation only half as important as product innovation (i.e. ratio=0.5), depending on the direction of change in coordination (i.e. increase or decrease).

The case study literature further provides concrete illustration for the above quantitative finding that firms in densely coordinated economies prefer process innovation. Tödtling and Kaufmann (2002), for example, examined patterns of innovation in Austria, one of the most strategically coordinated economies among OECD countries (ranked top and second on respectively the Hall–Gingerich and Hicks–Kenworthy indicator). Set in the province of Upper Austria, a region with 1.4 million inhabitants and located between Vienna and Munich, the study focused in particular on small to medium-sized enterprises (SMEs) that have less than a hundred employees, which make up 96 percent of firms in the province's main sector, manufacturing. Based on their survey of 204 firms in the region, the authors found that, in more than half of the SMEs, technological innovation is in the form of introducing "labor-saving technologies" (p. 24) leading to increases in productivity, in other words, process innovation. For example, among SMEs that receive various sources of public funding (such as the European Recovery Program and the Austrian Science Fund), 95.6 percent identified "improving productivity" or "flexibilization of production" as their motive for innovation, while only 68.9 percent identified "technological basis for new products" (percentages sum to more than 100 percent because a given firm may identify with both product and process innovation). Among firms that do not receive public funding for innovation, 100 percent identified with some form of process innovation, while only 26.1 percent identified with product innovation. The preference for process over product innovation is also reflected in Austrian firms' market strategies. As the authors found, SMEs in Upper Austria focus on increasing their share of existing product markets in the local region, rather than making entry into new product markets on the international stage.

Besides reinforcing evidence for more process innovation in more coordinated economies, qualitative findings from the case-study literature also help illuminate the causal mechanism underpinning this outcome: firms that specialize in process innovation are more willing to engage in relational sharing of technological resources than those that emphasize product innovation. For example, as a high-tech radical innovation sector, biotech is commonly regarded as an area of strength for liberal market economies such as the US. However, as Casper (2000) found, Germany has made a surprising breakthrough in the "platform technologies" segment of this sector, while the US continues to dominate in the "therapeutics" segment. Technological innovation in therapeutics is a classic example of product innovation, driven by the use of molecular biology and genetic engineering to develop new treatments. Platform technologies, by contrast, focus on process innovation, creating "enabling technologies" that "*rationalize* common molecular biology lab processes," "*automate* many aspects of the discovery process," and "*aid* in the quest to fully decode and understand the human genome," such as high-throughput combinatorial chemistry applications and genetic sequencing and modeling techniques (Casper 2000, p. 895, emphasis added). In other words, while the US maintains a stronghold on "product innovation" in biotechnologies, Germany has managed to carve out a "process innovation" foothold in this high-tech sector. Underlining the contrast between these two innovation strategies are two very different relational processes. As Casper pointed out, the development of new drugs in therapeutics is subject to considerable appropriability risks, because the innovator's profits will be dissipated by the access of other firms to its new product. As a result, innovating firms must manage their interaction in a way that *prevents* the substantial sharing of innovation rents with others, which requires a relatively formal arm's length relationship more commonly found in liberal market economies such as the US. While platform technologies can also be easily appropriated, attempted exclusion of others from the market is not a viable strategy, since process technologies such as DNA cloning and genome sequencing have now become widely available. Rather than monopoly rents, innovators focus on collecting productivity rents by improving cost efficiency and offering lower service prices to pharmaceutical companies and other major customers. As a result, rather than trying to exclude each other, multiple platform technology firms often enter the same market, and focus on process technologies that have broad, common, applications, in order to enjoy the positive learning externality from each other. Such reciprocal sharing of knowledge requires a deep relational setting that is more easily established in a strategically coordinated economy such as Germany.

A very similar logic is uncovered in Biggiero's (2002) study of another "uncharacteristically European" center of biomedical innovation, the Italian city of Mirandola, located in the densely networked Emilia-Romagna region of the Third Italy. While small indigenous firms in Mirandola were able to develop new drug products, they faced financial and market barriers in further process innovations that are needed to streamline the production process, an advantage typically held by larger international firms. Nevertheless, Mirandola was able to overcome this barrier by becoming a magnet for multinational corporations (MNCs) that specialize in biomedical process innovation. What is particularly revealing about MNCs' choice of Mirandola was that they were attracted not by the prospect of exploiting wage differentials with the rest of Europe, but instead by the densely networked social capital in the region, which they kept fully intact, as well as the entire local management structure. Casper and Whitley's (2004) study of Sweden's success in the software industry (another sector commonly associated with liberal market economies) identifies a similar causal mechanism. Sweden has Europe's largest concentration of publicly listed firms in a segment of the software sector known as "middleware," which, despite its name, is actually marked by highly radical innovation in technologies that "*improve the efficiency* by which different computing systems interfaced with communications networks" (p. 97, emphasis added). "Middleware" technologies, in other words, generate process innovation that improves the performance of existing information technology (IT) services and products. This forms a contrast to "standard software," an area of product innovation in the form of ready-for-consumer-use multimedia, entertainment, and graphics software programs, where the liberal market UK enjoys a comparative advantage. As the authors found, the "middleware" sector is characterized by highly complex technologies that require significant coordination across firms in standards, as well as resolution of collective action problems in knowledge generation. Herein lies the key to Sweden's breakthrough: in a manner not dissimilar to the German biomedical success, a dense, deeply tied, network of Swedish firms engage in the development and sharing of common "platform technologies" and technical standards, which serve as "club goods" for all "middleware" firms involved. As another example, while Germany's Baden-Württemberg region has succeeded in the "process" segment of the multimedia sector (hardware and research for multimedia production), the "product" segment (new media contents and applications) is underrepresented in the region. A key reason for this outcome, as Krauss and Wolf (2002) pointed out, is that the relational sharing institutions traditionally present in the German economy may undermine the very purpose of product innovation. For example, multimedia firms felt that Germany's various systems of technological sharing and transfer (such as the Steinbeis Foundation) have generated "few incentives to them" (p. 43).

Finally, evidence from the case-study literature can illustrate the core pattern of product and process innovation from an angle I cannot capture in my quantitative analysis: *subnational* variation in coordination. Asheim and Iskasen (2002), for example, pointed out a contrast between two types of innovative regions in Norway. The first one, represented by shipbuilding in Sunnmøre and mechanical engineering in Jæren, focuses on process innovation, especially in "incremental improvements on the shop floor" (p. 79), while the second one, represented by the electronics industry in Horten, focuses on product innovation, developing patented products and systems that are brand new to the market. Behind this contrast in innovation strategy lies a contrast in the relational nature of the regional economy. Firms in Sunnmøre and Jæren rely heavily on reciprocal cooperation mediated through local organizations, such as the Mechanical Engineering Association in the District of Ulstein and Maritime Nordvest. The dense interfirm networks in Jæren, in particular, bear strong resemblance to the high-trust Third Italy region, and Norwegian SMEs in this region are tied closely together through a local industry organization called TESA (TEknisk SAmarbeid), which supports technological sharing among firms. Drawing from these institutional resources for knowledge-sharing, the region has created one of Norway's most successful process technology firms, ABB Flexible Automation, which supplies around 70 percent of painting robot technologies to car manufacturers in Europe. In the electronics industry of Horten, by contrast, firms engage in more "incidental" cooperation, where innovators are temporarily brought together because of similar formal knowledge in a specific innovation project, rather than being embedded in a common local socioeconomic network. Furthermore, sometimes firms in Horten simply bypass relational building with other firms, and conduct R&D inside their own corporate subdivisions instead. A similar pattern of subnational variation in coordination can also be identified in Sweden. For example, Rafiqui's (2010) study of the furniture industry in Sweden showed that while this industry continues to remain highly innovative and successful in one region (Tibro), it has become all but extinct in another traditional stronghold (Virserum). Furniture is another case where process innovation (improving the craft and productivity of production) is more important than product innovation (introducing new products to the market). Why did Virserum wither while Tibro remains thriving? Historically, furniture making in Tibro had been a process of cooperation among multiple small firms, across the vertical chain of production, while Virserum had a more "conglomerate" pattern of production, where all stages of the vertical chain were produced within the same single, large, firm. This historical difference in industrial structure turned out to be crucial in shaping firms' incentives to invest in relational institutions for tacit technological knowledge-sharing, which are crucial for the ultimate success of process innovation. Because all innovation is in-house, firms in Virserum had long adopted a "go it alone" approach to furniture making, and refrained from cooperation in innovation. SMEs in Tibro, by contrast, relied heavily on the collective use of common factory platforms and machine-sharing, while different firms in the sharing network specialized in different stages of the common production chain. As a result, over time the more strategically coordinated Tibro region outcompeted the arms' length Virserum region in process innovation in the furniture industry.

Overall, both the chapter's core quantitative analysis and the supplementary evidence from the case-study literature suggest that strategic coordination affects the nature of technological innovation. The more strategically coordinated the economy, the more firms shift away from product to process innovation, and vice versa. In other words, despite possible synergies between process and product innovation, they thrive in opposite relational environments. By choosing differently between product and process innovation, liberal market and coordinated economies also face different outcomes in employment or productivity. Next, I draw out these implications in full.

Whose Innovation Is Better for Employment and Productivity?

Whose Innovation Creates More Jobs?

Different countries engage in different mixtures of product and process innovations. In order to determine whose innovations overall are more employment-friendly, the first task is to compare the employment effect of product and process innovation. As I will show, these two types of innovation have very different implications for employment: one creates additional jobs, while the other is labor-saving. Then, taking into consideration how countries differ in their relative emphasis on process over product innovation, we can gain a better understanding of how countries differ in the overall employment effect of innovation.

Although relatively new to the comparative political economy literature, the differing employment effect of product and process innovation is a familiar topic in the innovation literature (Edquist et al. 2001; Pianta 2005; Vivarelli et al. 1996). By introducing new lines of products, product innovation opens up new markets and creates new job opportunities. By contrast, process innovation raises productivity in the production process, allowing more products to be made faster, with less input and overhead. Because greater time and cost efficiency reduces the number of workers needed for creating a given value-added, process innovation saves labor and reduces employment.

H2: product innovation increases employment; process innovation reduces employment.

The causal processes implied by this hypothesis speak directly only to the specific impact on jobs *within* the innovating sector, but the total job impact *on the national level* is often of greater relevance and importance, to comparative scholars as well as policy-makers. When we move the inquiry beyond the micro, sector, level, the aggregate impact of innovation on employment may well become more moderate in size. In particular, when a sector/firm expands its jobs through product innovation, new product entry may also take markets away from other, incumbent, sectors/firms, and hence displace some of their jobs. Similarly, when a sector/firm sheds jobs through process innovation, some of these jobs may be absorbed by incumbent sectors/firms.

Because job gains (losses) in the innovating sector may be offset by losses (gains) in incumbent sectors, it is important to acknowledge at the outset that the ultimate, national-level, implication for employment will not be as stark as the predictions on the firm or sector level. Nevertheless, there are theoretical reasons to believe that there is a limit to the extent of such "mutual offsetting" in job gains and losses, and as a result, the national-level impact will be bounded strictly *away from zero*. Because much of the interfirm and intersector job migration reflects the competition for market between firms/sectors, I can draw from the rich literature on "monopolistic competition" (Hotelling 1929; Chamberlin 1937; Dixit and Stiglitz 1977; Gabsweiciz and Thisse 1986; Johnson and Myatt 2006) to address this issue. Product innovation, by definition, leads to a new product that is somewhat different from incumbents on the market. Product differentiation reduces product substitutability and relaxes price competition, allowing the entrant and incumbent to *coexist*, each capturing a different segment of the market (Sutton 1986; Raith 2003). Product differentiation may be *horizontal* (such as different design styles catering to different tastes) or *vertical* (such as premium and low-quality targeting respectively rich and poor consumers) (Shaked and Sutton 1987; Philips and Thisse 1982). In either case, both the entrant and the incumbent will remain in the market with a positive share (Lancaster 1982; Sutton 1986). Therefore, although the entrant has a monopoly on the *new product offering* (which allows it to earn novelty/monopoly rents through a price premium), it will not achieve a monopoly in *market share* (hence the literature's term "monopolistic competition" in reference to product differentiation). In other words, product differentiation prevents the entrant from completely substituting the incumbent's market. Therefore, there will be a limit to which new jobs from product innovation simply displace old jobs from incumbent sectors. For this reason, the net impact of product innovation on jobs should still remain *positive on the national level*.

Process innovation, by contrast, presents an opposite competitive environment: instead of differentiation in products, firms compete in production efficiency (and hence cost and pricing). With more substitutable products and fiercer price competition, a technologically efficient low-cost entrant can indeed displace an inefficient high-cost incumbent (Raith 2003). In other words, under process innovation, jobs are *not only* lost in the innovating sector (due to its greater productivity) *but also* in incumbent sectors (due to their disadvantage in technological efficiency). As a result, the net impact of process innovation on jobs should still remain *negative on the national level*.

As the discussion above indicates, although the employment impact of product and process innovation may be less stark on the national level, it will still be bounded *away from zero*, and therefore of empirical and practical interest to assess. In the following empirical analyses, I start with direct tests of hypothesis H2 on the sector level, taking advantage of the rich industry-level data in CIS. These sector-level findings will pave the ground for the testing, later in the chapter, of national-level implications for how innovation affects employment across varieties of capitalism.

Hypothesis H2 suggests that product innovation increases employment while process innovation reduces employment. This hypothesis has already been tested by various scholars of innovation (Brouwer et al. 1993; Antonucci and Pianta 2002; Vivarelli et al. 1996). Given the innovation literature's primary focus on technologies rather than the political economy of employment, these studies understandably did not control for various socioeconomic factors that may affect the level of employment. However, from the book's perspective of comparing capitalisms, it is important to parse out the effect of new technologies from the effect of other socioeconomic factors often noted in the comparative literature. I will address this issue further when I discuss the control variables.

Because the testing of hypothesis H2 relies on the differentiation of product from process innovation, I continue to use the indices for the intensity of product and process innovation constructed earlier from the CIS. As readers may recall, the index for product innovation intensity is an average of two source indicators each measuring one specific aspect of product improvement from innovation. Similarly, the index for process innovation is the average of two source indicators each capturing one dimension of improvement in the production process due to innovation. In the 2001 survey, for example, each dimension of improvement has three possible scores (one, two, and three for respectively "low," "medium," and "high"), and firms are asked to choose a category corresponding to their experience. I weighted each score with the percentage of firms choosing that score, and then summed the (weighted) scores as a measure for the extent of product (process) improvement in that specific dimension. As a result, the composite indices for both product and process innovation are continuous, and bounded between one and three. Product (process) innovation is at its maximum intensity when 100 percent of firms in a given sector choose the highest score (i.e. three) for product (process) improvement; conversely, product (process) innovation is minimum when 100 percent of firms choose the lowest score (i.e. one).

Because of its median position, the index value of "two" may be a useful benchmark for comparison in later interpretations of estimation results. The exact context for this median score may be diverse. For example, an index value of "two" may correspond to a situation where *half* of all firms report *low* intensity (i.e. a low score of "one") and *half* report *high* intensity (i.e. a high score of "three"). However, it may also correspond to a situation where *all* firms report *medium* intensity. These are, of course, only two of the many possible scenarios that lead to a mean score of "two" for firms across the sector. In other words, while an index value of "two" suggests that the product (process) innovation intensity across the sector *as a whole* is medium, it does not rule out the possibility that a particular individual firm in this sector may experience below- or above-medium intensity in product (process) innovation. Nevertheless, in reality, further distinction between these firm-level scenarios will not add substantively to the analysis, because data for both the dependent variable (employment) and core independent variables (product and process innovation) can only be disaggregated to the sector level, precluding any analysis on the level of the individual enterprise. For this reason, while I acknowledge that the experience of individual firms may vary, I interpret the median index value of "two" as a stylized situation, where 100 percent firms report medium intensity in product (process) improvement. This establishes consistency with my language of description for the boundary values of "one" or "three" (100 percent firms reporting high or low intensity), which facilitates the interpretation of empirical findings.

The dependent variable is employment. There are no direct questions about the employment effect of innovation in CIS, so data on employment have to be obtained elsewhere. There are long time series data on employment broken down by informative categories and widely available across a large number of OECD countries. However, the time frame for the employment data is restricted by the very limited time frame for the core independent variables: product and process innovation from CIS. As noted earlier, CIS is not a panel survey. Furthermore, the number of time points with available data (four waves) is much smaller than the number of countries (up to twelve). Because the data provides much less information from cross-time variation than cross-sectional variation, I analyze the data from each individual survey as a single cross-sectional analysis, and use findings from different waves as robustness checks against each other. For each survey year, I use the cross-sectional employment data from that year as the dependent variable. Because the core independent variables (product and process innovation) are measured on the sector level for each country, I also break down employment by sectors.

As an example, for the 2001 wave of the survey, the dependent variable is each country's employment in 2001, as a proportion of working age population (15–65 years of age), in sectors for which product (process) innovation intensity is measured in the survey. Data for cross-country sector-level employment rates is available from the OECD Structural Analysis database. Because the employment effect of innovation may not be instantaneous, I also run robustness checks where the dependent variable has a one-year lead (i.e. employment level one year after the survey's reference period), which yielded similar results. I report results based on the "no-lead" (contemporaneous with the start of the survey reference period) version of employment. For the analyses of the 2004, 2006, and 2008 surveys, I follow the same procedure, and use the cross-sectional sector-level employment rate as the dependent variable.

Besides the intensity of product and process innovation as the core independent variables, I also control for various other socioeconomic determinants of employment. I start by controlling for trade and capital openness. I measure trade openness as export plus imports, and capital openness as direct investment inflows, both in percentage of GDP and available from the World Development Indicators. I also control for features of the wage-setting process, by including not only union density (union membership as percentage of wage and salary workers) but also bargaining centralization. I measure bargaining centralization using Jelle Visser's five-point bargaining centralization index, which is based on Lane Kenworthy's (2001) classification of bargaining centralization. In this classification, a country is assigned a score of "five" if bargaining is economy-wide, "four" if there is mixed industry-wide and economy-wide bargaining, "three" if bargaining is industry-wide without regular pattern setting, "two" if there is mixed industry-wide and firm-level bargaining, and "one" if bargaining is fragmented at the firm level. Data for the union density and bargaining centralization variables are from Visser's Database on Institutional Characteristics of Trade Unions, Wage Setting, State Intervention, and Social Pacts.

I include both union density and bargaining centralization as controls because, as Bradley and Stephens (2007) suggest, these two variables capture important aspects of the debate between neoliberal and neocorporatist schools on the determinants of employment. The reservation wage theory in economics focuses on the price of labor as the main determinant of the employment level (OECD 1994, 1999, 2002a; Siebert 1997), implying that greater bargaining decentralization and weaker unions help reduce wage demands and hence increase employment by clearing the market. By contrast, the neocorporatist literature makes a more subtle argument about the effect of unions on employment (Iversen et al. 2000; Franzese 2002; Wallerstein 2008; Baccaro and Simoni 2010; Lundvall 2013). Drawing from Mancur Olson's (1965) insight on collective action, students of neocorporatism argue that centralized institutions of wage bargaining may actually facilitate wage discipline (and hence boost employment), because centralization prevents individual unions from freeriding on each other's efforts to exercise wage restraint. Seen in this perspective, the "power of labor" is detrimental to job creation only under the wrong combination: strong unions with decentralized bargaining. In other words, while one aspect of labor's bargaining power (union density) may indeed exert some upward pressure on wages and have a negative impact on employment, the other aspect (bargaining centralization) may enforce wage restraint and therefore make a positive contribution to employment. In light of this argument, labor strength and bargaining centralization may affect employment in opposite ways: union density can be detrimental to employment, but bargaining centralization boosts employment. Furthermore, I also include in the estimation social security taxes as percentage of GDP, based on the OECD's Revenue Statistics. While income taxes are borne by workers and can be used to fund a wide range of public goods and services (such as infrastructure, active labor market training, and public education), payroll taxes may have a very different impact on the labor market: they add directly to firms' cost of labor, and bind the funding to "passive" income transfers, such as unemployment benefits, pensions, and disability benefits. For this reason, heavy payroll taxes may also exert downward pressure on the level of employment in the economy.

Information from these control variables is, unfortunately, coarse in two ways. First, although both the dependent variable (employment) and the core independent variables (product and process innovation) can be broken down by sector, the control variables are only measured on the country level. Second, although all controls have long time series stretching back to the 1970s, the analysis of each CIS survey relies purely on cross-sectional variation. To fit into this cross-sectional framework, I construct a cross-sectional version of the controls, by taking each control's average value over the ten most recent years preceding the survey. Of course, these cross-sectional controls lose all the cross-time variation in the original data, but this is inevitable given the cross-sectional nature of the survey analysis.

These controls capture some important features of coordinated economies which, according to the neocorporatism literature, may affect employment (such as bargaining centralization). However, what if there are other unmeasured aspects of national-level coordination, which directly affects employment rather than indirectly through product and process innovation? As noted earlier, both employment and innovation are measured on the sector level. As a result, much of the variation in the core independent variables (product and process innovation) occurs between individual sectors on the subnational level. In other words, the sector-level setup of the analysis provides a relatively clean source of subnational variation in product (process) innovation, not confounded by differences in national-level coordination. I report the results of the SUR analysis in Table 3.2.

Comparing the point estimates for product versus process innovation from the 2001 survey, they are both statistically significant, and oppositely signed, consistent with my theoretical expectations: while product innovation has a positive effect on employment, the employment effect of process innovation is negative. Based on findings from the 2001 survey, a one-unit increase in a sector's product innovation intensity will raise that sector's employment rate by around 6.8 percent, while the same extent of increase in process innovation will lower employment by around 3.7 percent. An increment of "one unit" happens to correspond to the difference between the innovation intensity index's median value (two) and boundary values (one and three). In other words, when all firms in a sector report high as opposed to medium intensity in product (process) innovation, the sector's employment rate will rise (fall) by 6.8 percent (3.7 percent). The potential employment impact from shifting between product and process innovation is therefore quite notable: when all firms report high as opposed to low intensity of product (process) innovation, an additional 13.6 percent (7.4 percent) jobs will be added (lost) in that sector. Of course, in reality, it is unlikely to directly observe this upper bound on the employment effect of product and process innovation, because the difference in product and process innovation between most countries and sectors will be less drastic. Using a substantively more relevant metric, each standard deviation of increase in product innovation (equal to 0.52 units on the innovation intensity index), for example, will raise employment by 3.5 percent, while the same extent of increase in process innovation (0.42 units on the innovation intensity index) cuts employment by 1.6 percent. With all other variables at their mean, a sector where all firms report medium product innovation has an employment rate of 59 percent. A standard deviation more product innovation will increase employment to 62.5 percent, and two standard deviations will raise it to 65.5 percent. Similarly, if all firms report medium process

Table 3.2. The Effect of Product and Process Innovation on Employment, Twelve OECD Countries

Community Innovation Survey Wave	Third Wave (2001) Equation a	Third Wave (2001) Equation b	Fourth Wave (2004) Equation a	Fourth Wave (2004) Equation b	Fifth Wave (2006) Equation a	Fifth Wave (2006) Equation b	Sixth Wave (2008) Equation a	Sixth Wave (2008) Equation b
Product Innovation	6.78(2.91)***	6.84 (1.86)***	5.61(1.2)***	5.54(0.89)***	2.81(0.37)***	2.21(0.24)***	3.12(1.29)**	2.90(1.40)*
Process Innovation	−3.79(0.83)**	−3.71(0.65)***	−3.21(0.84)***	−3.75(0.80)***	−5.27(1.4)***	−4.30(0.98)***	−2.65(0.89)***	−3.40(1.3)**
Capital market openness	−0.41(0.19)*	−0.35(0.17)*	−1.01(0.35)**	−1.29(0.23)***	−0.87(0.23)***	−1.10(0.46)*	−0.95(0.19)***	−0.67(0.33)**
Trade openness	0.11(0.03)**	0.07(0.02)**	0.36(0.11)***	0.44(0.19)*	0.21(0.10)*	0.29(0.08)***	0.28(0.09)***	0.24(0.09)**
Social security taxes	−0.42(0.21)*	−0.21(0.19)	−0.69(0.35)	−0.81(0.35)*	0.65(0.59)	0.43(0.45)	−0.51(0.41)	−0.38(0.19)*
Bargaining centralization	−1.24(1.01)	0.01(0.84)	0.90(0.40)*	0.79(0.65)	0.67(0.16)***	0.56(0.23)**	0.89(43)*	0.80(0.37)**
Union density	0.08(0.05)	0.06(0.02)***	0.19(0.12)	0.21(0.18)	0.02(0.01)	0.02(0.01)	0.11(0.04)**	0.08(0.04)
Constant	12.51(0.97)***	−0.34(0.06)***	14.62(0.81)***	6.88(0.21)***	11.24(0.83)***	9.66(0.57)***	11.10(1.12)***	6.12(0.23)***
N	64	64	175	175	202	202	241	241
Parms	15	12	37	12	57	12	87	12
RMSE	3.10	2.69	4.14	2.51	3.30	3.35	2.20	3.20
p	0.00	0.00	0.00	0.00	0.00	0.00	0.00	0.00
R-squared	0.70	0.41	0.59	0.43	0.48	0.39	0.60	0.53

* $p<0.1$; ** $p<0.05$; *** $p<0.01$

innovation, the sector's employment rate will be 61.6 percent, with all other variables held at mean. One standard deviation (more process innovation) will reduce employment to 60 percent, and two more standard deviations will cut it to 57.9 percent.

This pattern is corroborated by similar findings from the 2004, 2006, and 2008 surveys in Table 3.2. For each survey, the point estimates for product and process innovation are statistically significant and oppositely signed. There is, therefore, a robust pattern where product innovation increases employment while process innovation is labor-saving. Depending on the survey year, each unit of increase in the intensity of product innovation will add somewhere between 2.21 percent and 6.84 percent to the sector's employment rate; by contrast, for the same increment of increase in process innovation, the sector's employment rate will be slashed by somewhere between 2.65 percent and 5.27 percent.

Product innovation creates jobs, but process innovation destroys them. These sector-level outcomes will also have national-level implications for how countries differ in the way innovation affects employment. As I noted earlier, although some of the job gains (losses) in the innovating sector may be offset by losses (gains) from incumbent sectors, there will be a limit to the extent of such mutual offsetting. In product innovation, differentiation on the product space prevents newly created jobs from fully displacing those in incumbent sectors. In process innovation, the lack of product differentiation allows the technologically efficient low-cost entrant to displace inefficient high-cost incumbents, causing job losses in both entrant and incumbent sectors. In other words, the national-level impact of innovation should be bounded strictly *away from zero*: positive for product innovation and negative for process innovation. The more coordinated the economy, the more process relative to product innovation, and the weaker the marginal contribution of innovation to employment. Innovation, in other words, interacts with coordination in its overall effect on employment.

> H3: *the more coordinated the economy, the smaller the marginal contribution from technological innovation to overall employment in the economy.*

The hypothesis H3 predicts an interaction effect: the marginal impact of innovation on employment is conditional on the strength of coordination. The stronger the coordination, the weaker the effect of innovation in boosting employment. Therefore, the interaction term should be negatively signed. Because this hypothesis examines the overall effect of innovation on employment, it is no longer necessary to separate product from process innovation. Without having to separate product from process innovation, it is now possible to work outside the cross-sectional constraint of CIS, and exploit instead some time series macro-level data on innovation. To establish the robustness of findings, I use both expenditure- and patent-based data to measure innovation. The former is measured as total enterprise R&D spending (in percentage of GDP), available from the OECD's Structural Analysis database; the latter is measured as the total number of patents (in thousands) granted by USPTO, available from Eurostat's Science and Technology (Patent Statistics) database. It is important to point out that R&D and patent are more likely to reflect innovation activities aimed at product differentiation than activities aimed at improving the production process (Comanor 1967; Gilbert and Newbery 1982). As a result, R&D and patent data is likely to miss out some process innovation, and since such innovation is more prominent in strategically coordinated economies, the data may underestimate these countries' lead over liberal market economies in process innovation. Since the employment outcome stipulated in hypothesis H3 is driven by countries' differences in process and product innovation, the underestimation of their differences in process innovation may bias the analysis *against* hypothesis H3.

I retain all the control variables used earlier in estimating how product and process innovation differ in their employment effect, but this time I use the controls in their full time series. With data availability for all variables considered, the analyses of hypothesis H3 cover fifteen countries[3] for 1980–2006. I report the results of the fixed-effect vector decomposition estimation (with Prais–Winsten adjustment for serial correlation) in Table 3.3.

I start the discussion with Model 1, which estimates the contemporaneous effect of R&D spending on employment. For this model, the first finding of interest is that the coefficient of the "main effect" (i.e. stand-alone) variable for R&D spending is significant and positively signed. In other words, when coordination is at its minimum (zero on the Hicks–Kenworthy index), innovation is job-creative rather than destructive: each increase in overall enterprise R&D spending by 1 percent of GDP translates into a 0.41 percent increase in the economy's employment rate. However, this is very much the upper bound on the potential job boon from investment in innovation. The strength of coordination in most countries is well above the minimum, and as the negatively signed interaction indicates, innovation's effect on employment is dragged down by coordination. The more coordinated the economy, the smaller the contribution of R&D spending to employment. As Model 1 shows, a one-unit increase in coordination will reduce the marginal contribution of R&D to the economy's employment rate by 0.66 percent.

To illustrate the interaction, Figure 3.1 plots the marginal effect of overall R&D spending on the economy's employment rate, conditional on the strength of coordination, with dashed lines marking the 95 percent confidence interval.

---

[3] These countries are Australia, Austria, Belgium, Canada, Denmark, Finland, France, Germany, Italy, Japan, the Netherlands, Norway, Sweden, the UK, and the US.

Table 3.3. The Effect of Innovation on Employment, Fifteen OECD Countries (1980–2006)

	Model 1	Model 2	Model 3	Model 4	Model 5	Model 6
Coordination*R&D	−0.66(0.10)***	n.a.	−0.62(0.12)***	n.a.	−0.41(0.12)***	n.a.
Coordination*Patents	n.a.	−1.13(0.35)***	n.a.	−2.26(0.53)***	n.a.	−0.51(0.21)**
Coordination	−1.65(0.74)*	−2.97(1.54)	1.59(0.69)*	5.87(4.6)	2.53(2.06)	4.31(3.27)
R&D	0.41(0.03)***	n.a.	0.37(0.04)***	n.a.	0.40(0.08)***	n.a.
Patents	n.a.	1.29(0.21)***	n.a.	1.99(0.25)***	n.a.	1.01(0.28)***
Capital market openness	−0.63(0.09)***	−0.98(05)***	−0.59(0.09)***	−1.22(0.14)***	−0.78(0.29)**	0.21(0.15)
Trade openness	1.12(0.22)***	0.67(0.28)**	0.98(0.13)***	0.33(0.10)***	1.42(0.52)**	0.68(0.26)*
Social security taxes	−0.13(0.10)	0.06(0.05)	−0.25(0.20)	0.00(0.007)	−0.91(0.30)**	−0.50(0.22)*
Union density	0.01(0.004)*	0.00(0.004)**	0.01(0.004)*	0.02(0.008)**	0.01(0.005)**	0.00(0.00)
Bargaining centralization	0.68(0.33)*	1.56(0.46)**	0.20(0.17)	2.25(0.87)**	1.35(0.64)*	0.25(0.12)
Constant	79.17(2.21)***	45.2(3.61)***	70.25(4.61)***	53.23(5.74)***	59.15(2.58)***	89.92(1.54)***
R-squared	0.36	0.38	0.34	0.41	0.30	0.29
N	298	256	286	269	298	242

* $p<0.1$; ** $p<0.05$; *** $p<0.01$

Whose Innovation Creates More Jobs?

[Figure: line chart with x-axis "Coordination" from 0.0 to 1.0 and y-axis "Effect of R&D on Employment" from -0.5 to 0.5. A solid downward-sloping line begins near 0.4 at coordination 0.0 and decreases to about -0.25 at coordination 1.0, with dashed confidence interval lines above and below.]

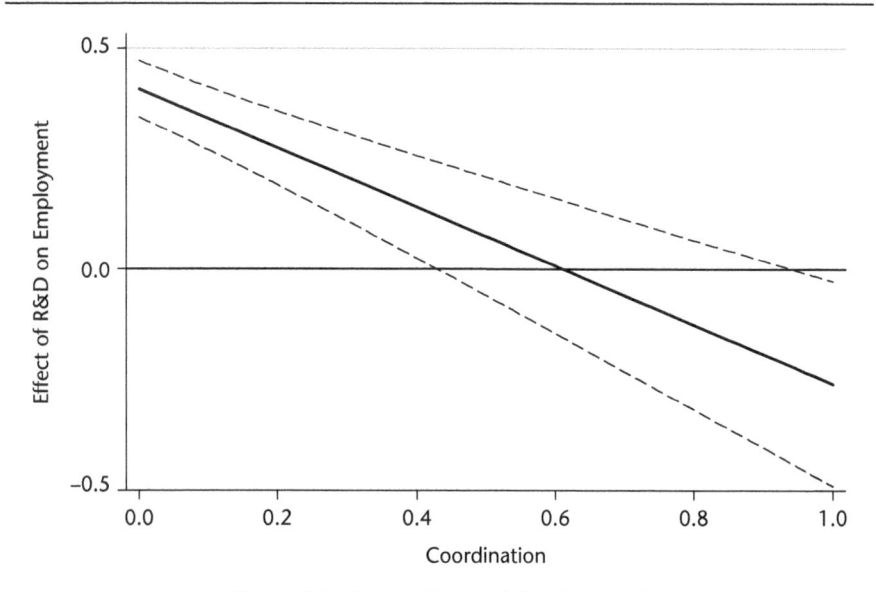

Figure 3.1. Innovation and Employment

As the graph shows clearly, the impact of R&D on the employment rate turns from positive to negative as coordination gains strength. When coordination is minimum at zero, an increase in R&D spending by 1 percent of GDP can boost the employment rate by just under 0.5 percent. This effect, of course, can also be directly recovered as the "main effect" coefficient for R&D in Model 1. This job-creative benefit of R&D, however, declines steadily as coordination becomes stronger, and drops out of significance after coordination grows to around 0.40 on the Hicks–Kenworthy index (roughly corresponding to Italy and France). When coordination reaches a very strong point (around 0.95 on the index, close to the level of Austria, Norway, and Sweden), the employment effect of innovation actually flips from positive to negative. In these heavily coordinated economies, in other words, innovation destroys jobs: each increase in R&D expenditure by 1 percent of GDP reduces the economy's employment rate by around 0.24 percent.

To better appreciate the substantive implications of such interaction, I follow the procedure by Kam and Franzese (2007) and calculate the marginal effects (and associated standard errors) of R&D specific to various countries. For example, while each 1 percent growth of R&D in the weakly coordinated UK (0.10 on the Hicks–Kenworthy index) will add another 0.34 percent to its employment rate, the same increment of increase in innovation in the Netherlands (0.58 on Hicks–Kenworthy) nudges Dutch employment upwards by a mere 0.03 percent, and with the standard error twice as large (0.07), this effect can be safely written off as insignificant. Similarly, strongly coordinated Japan

(0.77 on coordination) falls behind the liberal market US (0.02 on coordination) in creating jobs from R&D: while each 1 percent R&D increase in the US raises its employment rate by 0.38 percent, the same R&D increase in Japan actually ends up slashing employment by another 0.1 percent, although this job-destructive effect is not statistically significant (standard error 0.09). The contrast between liberal market and coordinated economies is even more starkly illustrated when one of Katzenstein's (1984) classic examples of strong coordination, Austria, is paired with the US. While innovation in the US adds 0.38 percent more to employment, each 1 percent R&D increase in Austria cuts the employment rate by around 0.25 percent, and with standard error at 0.11, this job-destruction effect actually clears significance at the 0.05 level. In other words, the difference in coordination strength between Austria and the US translates into an American job advantage of almost 0.7 percent of the labor force, for each 1 percent of GDP in R&D spending.

The comparative political economy literature has pointed out that a large sector of noninnovative low-skill jobs made an important contribution to Anglo-Saxon economies' higher employment than Europe (Iversen and Wren 1998). This chapter's findings imply that the "bad job" interpretation of the US job miracle is only part of the picture: liberal market economies are not only very successful in sustaining dead-end jobs, but also in using R&D to create new jobs. Similarly, continental European economies face high unemployment not only because of a rigid insider/outsider divide on the labor market (Rueda 2007), but also because their manufacturing firms engage in the wrong (from the perspective of job creation) kind of innovation: process innovation, which raises productivity but saves labor. In a similar vein, although the active integration of labor market outsiders (through active labor market policies, universal welfare states, and high-quality public education) has allowed Scandinavian countries to notably outperform continental Europe in service-sector employment (Thelen 2014; Steinmo 2010; Gornick and Meyers 2003; Iversen and Stephens 2008; Nelson and Stephens 2012), this success could have been even greater, had these heavily coordinated Nordic economies been able to shift from labor-saving process innovation to product innovation instead.

In lieu of R&D spending, Model 2 in Table 3.3 uses the total number of patents (in thousands) granted by USPTO to measure the extent of innovation in the economy. As the "main effect" coefficient for patents shows, when coordination is at its minimum (zero), each one thousand increase in the number of patents in the economy raises its employment rate by 1.29 percent. This, again, is in fact the very upper limit on the possible employment benefit from new patents. The overall contribution of innovation to employment will decline as coordination strengthens and drives the economy from (job-creating) product innovation to (job-destroying) process innovation.

Whose Innovation Creates More Jobs?

Reflecting this "drag on employment" caused by coordination, the point estimate for the interaction term is negatively signed and statistically significant. As Model 2 shows, a one-unit increase in the strength of coordination will reduce the marginal contribution of new patents to the economy's employment rate by 1.13 percent.

Based on Model 2, Figure 3.2 plots the marginal employment effect of patents, conditional on the strength of coordination. The positive effect of new patents on employment diminishes as coordination grows, and roughly after coordination reaches median strength (close to Italy and Switzerland), patents no longer have any statistically significant effect in job creation. Unlike R&D spending (Model 1) whose employment effect eventually flips over to negative when coordination becomes very dense, the confidence interval for patents remains widely straddled across zero even at maximum coordination. This implies that, when innovation is measured in patents, the "employment drag" from coordination is somewhat less serious: coordination may completely prevent patents from adding new jobs to the economy, but it does not aggravate the situation by turning patents into a source of net job loss.

While Models 1 and 2 in Table 3.3 examined the contemporaneous effect of innovation on employment, Models 3 and 4 consider employment one year after innovation, measured respectively in R&D spending and patents. As the negative interaction coefficients from these two models indicate, the core pattern of finding remains the same when we examine employment one year after innovation: R&D and new patents are less effective in raising

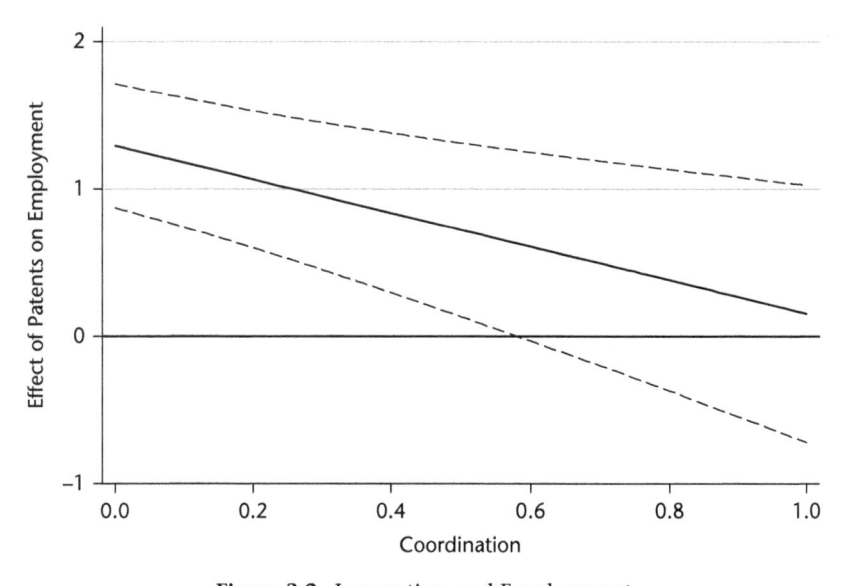

Figure 3.2. Innovation and Employment

employment when the economy is more coordinated. As a further robustness check, Model 5 reruns Model 1's analysis of how R&D affects employment, this time using Hall–Gingerich in lieu of Hicks–Kenworthy in measuring coordination. Model 6 turns to patents, re-estimating Model 2 with Hall–Gingerich as the alternative coordination measure. The finding again survives. In both Models 5 and 6, the interaction between innovation and coordination is signed correctly (negative), and clears significance at least on the 0.05 level. In other words, the more coordinated the economy, the less effective is technological innovation in increasing employment.

Having now established a general pattern of how coordination makes technological innovation less employment-friendly, now I draw out a few additional implications from this finding, by breaking innovation or employment down into some substantively meaningful subcategories. As I noted earlier in this chapter, the product/process innovation cleavage cuts across, rather than overlaps with, the well-known distinction in the literature between radical and incremental innovation. A product innovation may lead to either considerably different products or marginally better products. Similarly, a process innovation may be relatively drastic or incremental. The more radical the product innovation, the newer the line of products and markets it opens up, and the more opportunities for new job creation. The same goes for process innovation: the more radical the improvement in productivity, the more products can be made faster with less input, and the more jobs are eliminated through improved time and cost efficiency. In other words, the contrast between new products and new production processes in their employment effect should be sharper in radical than incremental innovation. As a result, by driving the economy from product to process innovation, the negative impact of coordination on employment should also be larger in radical than incremental innovation.

In order to explore this possibility, I run two separate estimations for the employment effect of innovation, respectively in radical and incremental technologies. To distinguish between radical and incremental technologies, I continue to follow the OECD guideline in classifying the extent of innovation, which I discussed earlier in Chapter 2 (Hatzichronoglou 1997). Using three criteria (production, value-added, and technologies embodied in the R&D process), the OECD identifies the following sectors with the most drastic extent of innovation: all those with two-digit NACE being 30, 32, 33, 64, 72, or 73, plus two three-digit sectors 24.4 and 35.3. Industries in this category range from computers, communications, and aerospace to pharmaceuticals. I aggregate R&D spending across these NACE digits (NACE-based R&D data are available from the OECD Structural Analysis database) to create a measure of overall R&D spending in radical technologies; for incremental technologies, I aggregate R&D spending across the

Whose Innovation Creates More Jobs?

Table 3.4. The Effect of Innovation on Employment, Additional Analysis

	Model 1	Model 2	Model 3	Model 4
	Radical Technologies	Incremental Technologies	Exposed Labor Force	Core Labor Force
Coordination*R&D expenditure	−2.53(0.42)***	−1.02(0.19)***	−1.30(0.09)***	−0.53(0.12)***
Coordination	−1.44(1.18)	2.61(1.15)*	0.05(0.04)	1.36(0.84)
R&D	1.66(0.27)***	0.74(0.06)***	0.70(0.04)***	0.43(0.05)***
Capital market openness	−0.42(0.12)***	−0.88(0.16)***	−0.65(0.09)***	−1.21(0.35)***
Trade openness	0.85(23)***	1.65(0.75)*	0.37(0.17)*	0.95(0.41)*
Social security taxes	0.02(0.01)	−0.04(0.02)*	−0.05(0.01)***	−0.08(0.05)
Union density	0.00(0.007)	0.02(0.007)**	0.10(0.04)**	0.00(0.01)
Bargaining centralization	0.20(0.12)	0.94(0.40)*	0.06(0.04)	0.51(23)*
Constant	62.54(2.31)***	81.41(1.65)***	73.14(2.98)***	75.29(6.55)***
R-squared	0.33	0.31	0.32	0.49
N	98	134	156	195

* $p<0.1$; ** $p<0.05$; *** $p<0.01$

remaining NACE digits. The results for these two analyses are reported in Models 1 and 2 in Table 3.4. Aside from breaking down innovation into two categories (radical and incremental), these two models retain all other variables used earlier in estimating the effect of innovation on employment in Table 3.3.

The coefficient for the interaction term is statistically significant and correctly signed (negative) for both radical and incremental technologies. In other words, for both types of technologies, coordination makes innovation less effective in creating employment. Where is this negative employment impact of coordination stronger? The point estimate for the interaction in radical innovation is about twice the size for its incremental counterpart, which implies that the negative conditioning effect of coordination is especially sharp when innovations are cutting-edge.

This pattern may be more easily observed in Figure 3.3, where the marginal employment effects of R&D in radical and incremental technologies are plotted (respectively in dashed and solid line) against coordination. Although the difference is not very dramatic, the slope for radical R&D is visibly steeper than the slope for incremental R&D, which implies a stronger interaction effect for radical innovation. On the other hand, radical R&D has a wider confidence interval, which implies that its interaction is estimated with less precision than incremental R&D. The relative positioning of the two confidence intervals sheds further light on where the two plots are genuinely distinguishable from each other. The marginal effect of radical technologies on employment is different from incremental technologies when either (1) one confidence interval straddles across the zero line while the other does not, or (2) they do not overlap. Seen from this perspective, it is clear that the

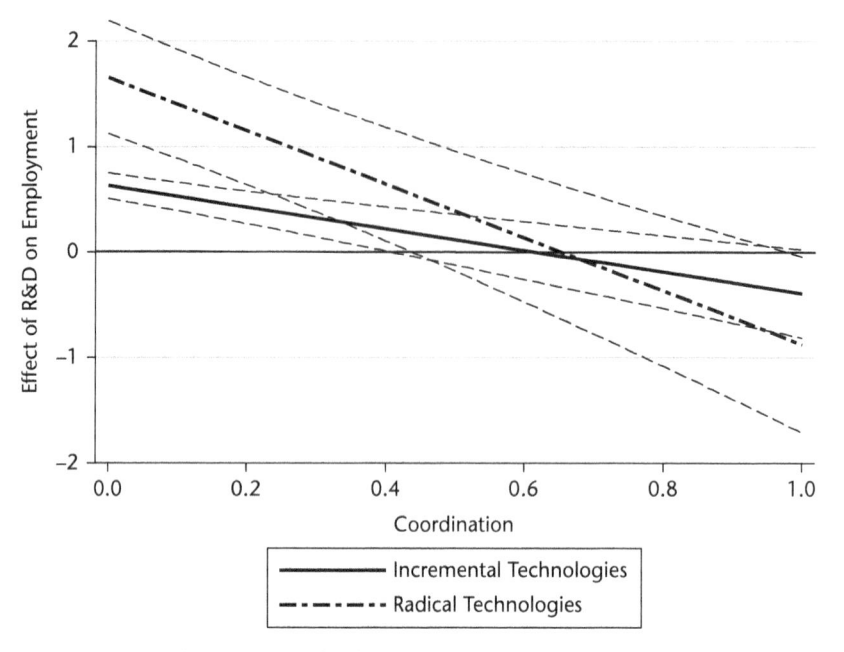

Figure 3.3. Radical vs. Incremental Technologies

difference between radical and incremental innovation is mainly in the lower ranges of coordination (below 0.2 on the Hicks–Kenworthy index, which neatly encompasses all Anglo-Saxon countries). In this region, weak coordination makes R&D more effective in job creation, and this effect is notably sharper when R&D is radical. Take a case of minimum coordination (0.0 on Hicks–Kenworthy, close to the US level) for example, each 1 percent R&D increase in incremental technologies will add 0.74 percent to the economy's employment rate, but the same R&D increase in radical technologies can produce more than twice the lift for employment (1.66 percent).

Having examined the employment effect from different slices of innovation, I turn to jobs and explore how the effect of innovation may differ for different slices of employment. Although all workers will be directly or indirectly affected by the introduction of new technologies, the elderly and the young may bear the brunt of the impact. The elderly may be the first in line when process innovation eliminates jobs, not only because their prospect for further training is limited, but also because they make attractive candidates when firms attempt to externalize the cost of downsizing onto the social security system through schemes such as early retirement and disability benefits (Ebbinghaus 2006; Huo 2009). The same vulnerability may also hit the young, especially when rigid employment protection creates an insider/outsider divide in the labor market that prevents firms from quickly dismissing the established core workforce (Rueda 2007). At the same time, however, the young may also be especially advantaged in the new job opportunities created by product innovation, not only because younger cohorts stay in school longer and attain more advanced academic education, but also because their relative youth permits a longer horizon for investment in skill training (Oster et al. 2013). As a result, workers in the younger and older segments of the labor force may be especially exposed to the creation and termination of jobs during technological innovation. In contrast to these "outer layers" of the workforce, the "inner core" of prime-age workers may be more resistant to the employment turbulence from new technologies.

In order to explore this possibility, I run two separate estimations of how innovation interacts with coordination in affecting employment, respectively for the "exposed" and "core" workforce. The former corresponds to the younger (15–24 years of age) and elderly (60 and above) segments of the labor force combined, while the latter refers to workers of prime working age (25–60). I obtain the average employment rate for both groups from the OECD's Labor Force Statistics and Labor Market Statistics. The results for these two analyses are reported in Models 3 and 4 in Table 3.4. The coefficient for the interaction term is statistically significant and correctly signed (negative) for both the exposed and the core labor force. In other words, coordinated capitalism makes innovation less effective in raising employment for both segments of the labor force. However, the harm done to the exposed workforce is almost 2.5 times larger than the core workforce: while the latter's point estimate for the interaction is –0.53, the former has a point estimate of –1.3. Again, the contrast between the core and the exposed outer layers of the labor force may be more effectively illustrated graphically. In Figure 3.4, the marginal effect of R&D on employment in the exposed and core workforce is plotted (respectively in dashed and solid line) against coordination.

Although the difference is not very dramatic, the downward slope for employment in the exposed workforce is visibly steeper than the slope for the core workforce, which indicates that the negative impact of coordination is stronger on the exposed workforce. Judging from the relative positioning of the two confidence intervals, there are now *two* regions on the Hicks–Kenworthy coordination index where the two plots are genuinely distinguishable from each other. First, when coordination is very weak (below 0.05 on the Hicks–Kenworthy index), the job-creating benefit of innovation is larger for the exposed workforce than their counterpart in the core. At the minimum level of coordination (close to the US), for example, each 1 percent of GDP in R&D adds 0.43 percent to the employment rate for the core workforce; for exposed workers, by comparison, the positive effect on jobs is stronger by more than 50 percent, raising the employment rate by 0.7 percent. Second, when coordination is sufficiently strong (above 0.65 on the Hicks–Kenworthy index,

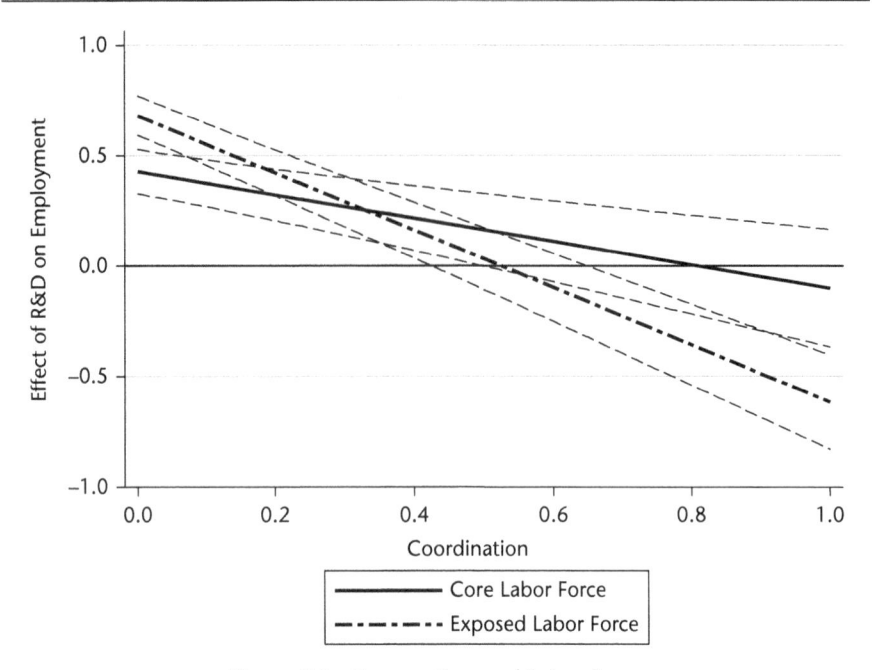

Figure 3.4. Core vs. Exposed Labor Force

which covers Austria, all Nordic countries, as well as Germany and Japan), the difference between the core and exposed workforce re-emerges, this time reflected in the loss, rather than the gain, of jobs from new technologies. While the confidence interval for the core workforce straddles *across* zero, it stays *below* zero for exposed workers. For example, with Swedish, Austrian, or Norwegian strength of coordination (close to 0.95 on the index), each 1 percent of GDP in R&D reduces employment in the core workforce by slightly under 0.1 percent, and the effect is statistically indistinguishable from zero. By contrast, the same increment of increase in innovation has a statistically significant impact on exposed workers, slashing their employment rate by almost 0.5 percent. The somewhat surprising finding that exposed workers in Nordic countries are especially vulnerable (to technological innovation) implies again that these countries' remarkable achievement in reintegrating labor market outsiders (through active labor market policies and high-quality public education) could have been even greater, had they been able to shift away from labor-shedding process innovation to product innovation instead. Such a shift, however, will be difficult for these cooperation-based economies, because relational sharing may *eliminate* the very novelty rents that are generated from product innovation.

This section has provided a detailed account of how the employment effect of innovation differs across varieties of capitalism. The less coordinated the economy, the more firms focus on product innovation, and the more effective is innovation in creating employment. There is, of course, another side to the story of how technological innovation affects work. When new technologies reduce the number of jobs needed for production, work becomes more productive. As I show next, precisely because productivity and the room for job creation may be negatively related, the employment disadvantage of coordinated capitalism during technological innovation may translate into an advantage in productivity. In other words, just as innovation in liberal market capitalism has an *extensive-growth* effect (expanding product lines and job opportunities), innovation in coordinated capitalism has an *intensive-growth effect* (deepening productivity and sophistication in the making of existing products).

Whose Innovation Deepens Productivity?

As shown earlier, the strong focus on process innovation has a negative impact on employment in strategically coordinated economies. The impact on the productivity of the workforce, however, may be quite different. Better and more efficient processes of production should increase the productivity of the workforce, because improvements such as smaller overhead, quicker turnover, and more flexible supply-chain coordination all enable workers to produce more output faster with lower cost and less input (OECD 1996a; Griliches 2000; Greenan et al. 2002). The more coordinated the economy, the more process relative to product innovation, and the more effective is innovation in lifting productivity in the economy. Innovation, in other words, interacts with coordination in its effect on productivity.

> H4: *the more coordinated the economy, the larger the marginal contribution from technological innovation to productivity.*

The hypothesis H4 predicts an interaction effect: the marginal effect of innovation on productivity is conditional on the strength of coordination. The stronger the coordination, the larger the overall effect of innovation in raising productivity in the economy. The coefficient for the interaction, in other words, should be positively signed. For aggregate productivity in the economy as the dependent variable, I adopt two different measures. The first one focuses specifically on labor productivity (as opposed to nonlabor factors of production). It calculates productivity as a "level" (i.e. output per unit of input), strictly following the OECD manual for measuring productivity detailed below. The second measure is broader in scope, considering aggregate productivity from all factors of production, labor or not (i.e. total factor productivity). It is measured as a "change" (i.e. growth in total factor productivity).

I start with the "level" measure of labor productivity. Its "output-per-unit-of-input" method of calculation is based on guidelines from the OECD (2001) manual *Measuring Productivity: Measurement of Aggregate and Industry-Level Productivity Growth*. In the manual, productivity is defined as the "ratio of a volume measure of output to a volume measure of input use." To calculate labor productivity defined as such by the manual, one needs separate data for the numerator (aggregate output) and denominator (aggregate labor input). For the former, I use total GDP in US dollars (adjusted for purchasing power parity). For the latter, I multiply the raw number of persons employed with annual working hours per worker, which range between 1,755 and 2,228 hours across advanced industrialized countries (working-time data are available from the OECD Statistics Compendium-Labor Market Database). Weighting employment by working time is important because, given the same size of the workforce, the actual amount of labor input is smaller in countries where working hours are short (such as in continental Europe) than where hours are long (such as in the US) (Boeri et al. 2008). By dividing aggregate economic output with hours-weighted aggregate labor input, I derive an annual "level" measure of labor productivity, calculated as economic output in US dollars per worker per hour.

The second measure of productivity is an annual "change" measure: the growth in total factor productivity. Rather than calculating an output/input ratio directly from empirical data, total factor productivity growth is derived instead as a residual, from the estimation of how factors of input contribute to economic growth. In other words, it is the component of economic growth that cannot be explained by the accumulation of input, broadly defined (labor and capital, including land). This calculation methodology stems from Robert Solow's (1957) treatment of economic growth based on the Cobb–Douglas function $Y = K^\alpha \times (AL)^{1-\alpha}$, where Y is the value of total output produced, K and L are respectively capital and labor input, α represents output elasticity, and A is total factual productivity, all for year t. Also known as the "Solow residual," total factor productivity growth is derived from this equation as follows. First, the equation is differentiated with respect to t, so that the size of annual growth in output ΔY is linked to the size of annual growth in inputs ΔK and ΔL, plus growth in total factor productivity ΔA, each with its own slope coefficient. Second, partial derivatives for K, L, and A are obtained as functions of output elasticity α, and then substituted into the above equation for the respective slope coefficients. Third, rearranging terms in the equation to leave only ΔA on the left hand side, the growth of total factor productivity can then be solved directly, based on the values of ΔK, ΔL, and α, all of which can be obtained from empirical data.

The growth of total factor productivity is commonly understood to represent both advances in technologies and increases in efficiency. The link between this productivity measure and new technologies is so well established that total factor productivity is sometimes regarded as another crude measure for the economy's long-term technological progress (Steil et al. 2002; Dollar and Wolff 1993; OECD 1996a). By directly estimating the extent to which R&D or patents contribute to the Solow residual, this chapter provides a more precise understanding of the extent to which increases in total factor productivity can be attributed to improvements in technologies, and more importantly, how the strength of this technology–productivity nexus differs across varieties of capitalism.

For a broad cross-section of OECD countries for 1980–2004, the Groningen Growth and Development Center (GGCD)'s Total Economic Growth Accounting Database provides annual data on total factor productivity growth. Because it is calculated as a residual, the measure itself is unit-free, and its numbers lack an intuitive interpretation. To draw out the substantive implications of productivity growth more intuitively, I first-difference the Groningen measure and then express each year's observation as a percentage change from last year. This way, each year's observation measures the extent of *acceleration* in productivity growth since last year. While this change-based measure of productivity is expressed as percentage increase in the speed of productivity growth, the earlier level-based measure of productivity is expressed as economic output in US dollars, per worker per hour.

I also include the following controls that may affect productivity in the economy independent of technological innovation. First, as studies of productivity growth point out, even in the context of constant technology (no innovation), labor productivity can still be raised by either simply increasing human capital endowment or investment in new machinery and equipment, even if such effect may be limited in the absence of parallel progress in production technologies (Dollar and Wolff 1993; Jorgenson 1995; Landau et al. 1996). To measure the overall human capital quality of the workforce as a whole, I control for the average years of schooling in the population aged 15 and over, which directly captures the central tendency in the labor force's human capital attainment. The data for this control is from the Barro–Lee (2013) Data Set of Educational Attainment. The Barro–Lee data is five-yearly, and given the generally upward secular trend in educational attainment over time, annual values are interpolated between each two five-yearly time points. For investment in machinery and equipment, I use fixed capital formation as percentage of GDP, available from OECD's Structural Analysis database subsection STAN Industry.

As noted earlier in the discussion of the Solow residual, allocative efficiency of economic resources may also contribute to productivity growth. From this perspective, a large public sector may potentially distort the allocation of resources and suppress economic productivity, as a result of the fact that, unlike private sector entrepreneurs, public sector agents face competing redistributive and/or bureaucratic objectives in the allocation of economic resources (Okun 1975; Kornai 1980). However, as Castles and Dowrick (1990) point out, different facets of the public sector may affect productivity and economic growth in different ways, depending on whether public resources are consumed by the state and its constituents, or channeled indirectly into human capital. As a result, I control for the size of the public sector from three different angles: general government consumption expenditure, social security transfers, and public expenditure on health, all calculated as percentages of GDP, from the World Bank's World Development Indicator, OECD's Social Expenditure Statistics and Health Statistics.

Although state expenditure on public health may interfere with allocative efficiency because of its need to meet various redistributive demands, this negative impact on economic productivity may be offset by the fact that public health itself boosts the quality of human capital. Social security expenditure may also have potentially conflicting effects on productivity. On one hand, generous social security may encourage unions to segment the workforce, resulting in labor market outsiders who are denied access to deep training and social protection (Rueda 2005, 2007; Palier and Thelen 2010). On the other hand, generous provision of income protection may help prevent the depreciation of productivity, which results from "forced matches" on the labor market. For example, high but reasonably short-term unemployment benefits may protect the unemployed from being forced, by financial pressure, to hastily enter a job that does not adequately preserve their existing skills (DiPrete 2002; Gangl 2004). Similarly, sick pay protects workers from being forced to return to work before complete recovery, which helps avoid the suboptimal utilization (and hence deterioration) of human capital while on the job (Huo et al. 2008).

As the IMF's *Government Finance Statistics Manual* (2001) explains, social security and government consumption expenditure are entirely different in the nature of spending. While the former is an example of "unrequited" transfer (not in compensation for work performed for the government), the latter represents "requited" transfer instead, i.e. state sector wage compensation plus costs charged for governments' use of goods and services from the private sector. In other words, consumption expenditure portrays the public sector with a special emphasis on the role of state-owned enterprises and industries. As a result, government consumption expenditure may play an especially important role in inhibiting economic productivity, via the channel of soft budget constraint and its second-order effects (shortage of resources, "forced substitution" using inferior resources, and disincentives for competition, all of which are typical pathologies from public ownership) (Kornai 1980, 1992).

With data availability for all variables considered, the analyses cover sixteen OECD countries[4] for 1980–2006 (when the dependent variable for productivity is measured as the output/input ratio), and fourteen OECD countries[5] for 1980–2004 (when the dependent variable is measured as growth in total factor productivity). I report the results of the (Prais–Winsten) fixed effects vector decomposition analysis in Table 3.5. Similar to the chapter's earlier analysis of employment, I used either contemporaneous or one-year-lead productivity as the dependent variable. Because estimation results from these two approaches are not substantively different, in Table 3.5 I report findings for contemporaneous productivity as the dependent variable.

I start the discussion with Model 1, which measures innovation through total private enterprise R&D expenditure, and measures labor productivity using the OECD's output/input ratio method. While the main parameter of interest is the interaction between R&D and coordination, the "main effect" variable for R&D provides a first cut into the pattern of findings. The point estimate for this variable is significant and positively signed. In other words, when coordination is at its minimum (zero on the Hicks–Kenworthy index), each 1 percent of GDP increase in R&D spending raises the level of workforce productivity by $6.54 per worker per hour. This, however, is only the lower bound on the potential contribution of R&D to labor productivity. As the statistically significant and positively signed interaction indicates, the more coordinated the economy, the stronger the marginal effect of technological innovation in raising workforce productivity.

As Model 1 in Table 3.5 shows, a one-unit increase in the strength of coordination will raise the marginal contribution of R&D spending to the economy's workforce productivity by $8.64 per worker per hour, which is as much as 22 percent of the average labor productivity ($39) across the countries in the analysis. Even a more modest one standard deviation (0.31) increase on the Hicks–Kenworthy index can add almost $3 to R&D's productivity effect for each man-hour. To illustrate this interaction, Figure 3.5 plots the marginal effect of R&D spending on the level of labor productivity, conditional on the strength of coordination.

As the graph shows clearly, the positive impact of R&D on worker productivity increases steadily as coordination gains in strength. When coordination is minimum at zero, 1 percent of GDP increase in R&D can raise productivity by just over $6 per worker per hour. This effect also corresponds to the "main effect" coefficient for R&D in Model 1. As the economy becomes

---

[4] These countries are Australia, Austria, Belgium, Canada, Denmark, Finland, France, Germany, Ireland, Italy, Japan, the Netherlands, Norway, Sweden, the UK, and the US.
[5] These countries are Australia, Austria, Canada, Denmark, Finland, France, Germany, Italy, Japan, the Netherlands, Norway, Sweden, the UK, and the US.

Table 3.5. The Effect of Innovation on Economic Productivity

	Annual Workforce Productivity as Output/Input Ratio (1980–2006), Sixteen Countries					Annual Total Factor Productivity Growth (1980–2004), Fourteen Countries				
	(1)	(2)	(3)	(4)	(5)	(6)	(7)	(8)	(9)	(10)
Coordination *R&D	8.64(1.36)***	n.a.	n.a.	3.37(0.95)***	n.a.	0.47(0.08)***	n.a.	n.a.	0.89(0.16)***	n.a.
Coordination* Applied R&D	n.a.	6.15(1.59)***	n.a.	n.a.	n.a.	n.a.	0.73(0.09)***	n.a.	n.a.	n.a.
Coordination* Patents	n.a.	n.a.	19.02(6.25)***	n.a.	10.04(3.5)***	n.a.	n.a.	2.65(.47)***	n.a.	0.91(.09)***
Coordination	3.67(0.76)***	3.10(0.35)***	4.59(1.62)***	1.18(0.36)***	n.a.	0.18(0.04)***	0.10(0.01)***	0.15(0.01)***	0.10(0.03)***	n.a.
R&D	6.54(0.43)***	n.a.	n.a.	6.67(0.69)***	n.a.	0.13(0.02)***	n.a.	n.a.	0.16(0.05)***	n.a.
Applied R&D	n.a.	4.52(0.19)***	n.a.	n.a.	n.a.	n.a.	0.21(0.08)***	n.a.	n.a.	n.a.
Patents	n.a.	n.a.	23.36(5.48)***	n.a.	14.01(0.3)***	n.a.	n.a.	4.51(0.99)***	n.a.	2.57(0.78)***
Years of schooling	9.81(1.63)***	7.68(1.83)***	9.22(2.30)***	11.78(1.75)***	5.70(1.74)***	0.10(0.01)***	0.11(0.01)***	0.19(0.04)***	0.12(0.02)***	0.04(0.00)***
Fixed capital	1.05(0.35)***	1.20(0.38)***	1.18(0.34)***	0.82(0.33)**	1.47(0.39)***	0.65(0.05)***	0.61(0.18)***	0.53(0.15)***	0.47(0.07)***	0.72(0.05)***
Public health	2.30(0.75)***	2.11(0.79)***	2.29(0.75)***	3.21(0.65)***	1.28(0.69)***	0.04(0.00)***	0.04(0.00)***	0.01(0.00)***	0.03(0.00)***	0.04(0.00)***
Social security	−1.65(0.43)***	−1.98(0.65)***	2.56(1.98)	−1.15(0.35)***	−1.93(0.57)***	0.14(0.05)***	0.16(0.04)***	0.18(0.04)***	0.19(0.05)***	0.09(0.05)***
Consumption expenditure	−1.25(0.87)	−1.20(0.87)	3.21(1.59)*	−0.98(0.94)	−1.45(0.68)*	0.11(0.04)**	0.11(0.05)**	0.12(0.05)**	0.08(0.00)***	0.11(0.01)***
R-squared	0.21	0.23	0.31	0.17	0.19	0.54	0.51	0.46	0.39	0.43
Constant	−22.1(7.0)***	−15.4(6.2)**	2.25(0.59)***	−29.6(6.6)***	1.78(0.35)***	−2.36(0.2)***	−2.98(0.3)***	3.25(1.78)	−1.86(0.28)***	−2.21(0.32)***
N	260	194	205	260	205	225	164	186	225	186

* p<0.1; ** p<0.05; *** p<0.01

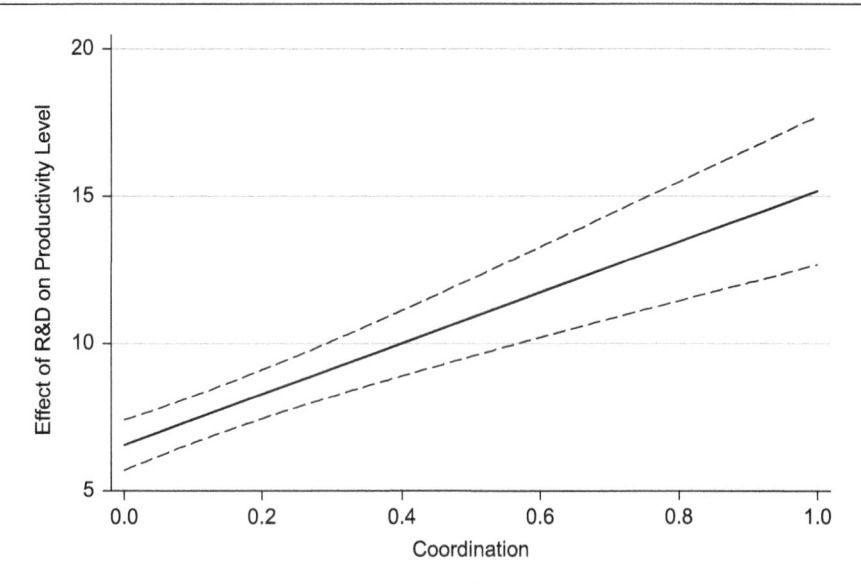

Figure 3.5. Innovation and Productivity Level

more coordinated, the productivity-enhancing impact of R&D becomes progressively stronger: at median coordination (0.5 on the Hicks–Kenworthy index, between Switzerland and Italy), each increment of R&D spending can already lift productivity by up to $11, and by the point of maximum coordination, the same increment is sufficient to add $15, almost three times the power when coordination is absent.

To discuss the substantive implications of such interaction, I calculate the marginal productivity effect of R&D specific to various countries. For example, while each 1 percent growth of R&D in weakly coordinated Canada (0.04 on the Hicks–Kenworthy coordination index) will add $6.9 to the hourly economic output per worker, the same R&D increase in strongly coordinated Denmark (0.72 on coordination) has more than twice the impact, lifting the productivity of Danish workers by $13.5. Similarly, highly coordinated Japan (0.77 on the Hicks–Kenworthy index) is ahead of the weakly coordinated UK (0.10 on Hicks–Kenworthy) in raising workforce productivity through R&D: while 1 percent growth of R&D raises British productivity by $7.1 per worker and hour, it adds twice as much ($14.2) to the output created by the Japanese worker. The productivity advantage of strategically coordinated capitalism is even more starkly demonstrated when the top of the coordination league (Sweden, 0.97 on the Hicks–Kenworthy index) is compared with the bottom (US, 0.02 on Hicks–Kenworthy): while one increment of R&D lifts the hourly productivity of the American worker by about $6.61, the productivity impact is almost three times larger for the Swedish worker ($15). In other words, the difference in coordination strength between Sweden and the US gives the former an innovation-based productivity advantage of about $8.39 per worker-hour, which is almost a fifth of the average workforce productivity across the OECD.

The OECD's (2002b) manual for measuring R&D expenditure (the *Frascati Manual*) distinguishes R&D spending by the extent to which the innovation leads to generic developments in fundamental scientific principles or practical application in commercial and industrial activities. While R&D oriented towards the former is defined as "basic research," R&D with greater emphasis on the latter is defined by the *Frascati Manual* as "applied" or "experimental" research. While Model 1 measured innovation as total R&D spending, Model 2 focuses specifically on R&D spending in applied and experimental R&D. Data for this subcategory of R&D spending is from the OECD Research and Development Statistics. Because applied and experimental research is more closely tied to the actual production process than basic research, the positive effect of coordination on productivity may be even stronger when we focus on applied research. However, comparing the interaction coefficients between Models 1 and 2 in Table 3.5, the actual difference turned out to be slight, and in any case, in the opposite direction (smaller for applied R&D than for overall R&D).

In lieu of R&D spending, Model 3 in Table 3.5 uses the total number of patents (in thousands) granted by USPTO to measure the extent of innovation in the economy. The statistically significant and positively signed interaction in Model 3 indicates that the findings are qualitatively unchanged: the stronger coordination, the stronger the impact of new patents on workforce productivity. As a further robustness check, Model 4 reruns Model 1's analysis of R&D and productivity, this time using Hall–Gingerich in lieu of Hicks–Kenworthy in measuring coordination. Model 5 turns to patents, re-estimating Model 3 with Hall-Gingerich as the coordination measure. Again, the core pattern of finding remains: the interaction between innovation and coordination is positive, and significant at the 0.01 level. Given the substantive similarity of findings, I omit separate graphs for Models 2 through 5.

While Models 1 through 5 examined the effect of innovation on the level of labor productivity, Models 6 through 10 turn to the growth of total factor productivity (i.e. the Solow residual). First, it is important to reiterate that, in order to avoid the unintuitive interpretation of raw Solow residual figures, the dependent variable is the Solow residual first-differenced and expressed as percentage change from previous year. In other words, it is the extent of *acceleration* in productivity growth since last year. As a result, a positive (negative) coefficient for innovation means that it causes the *growth* of total factor productivity to *accelerate* (*decelerate*).

Model 6 measures the extent of innovation using R&D spending. Again, I start by discussing the *lower* bound on innovation's effect on productivity growth (i.e. when coordination is minimum at zero). Then, I illustrate how

Whose Innovation Creates More Jobs?

this productivity effect intensifies as coordination gains strength. As the "main effect" coefficient for R&D indicates, when coordination is zero, each 1 percent of GDP increase in R&D accelerates total factor productivity growth by about 0.13 percent relative to last year, which happens to be around a tenth of the average acceleration rate in productivity growth (1.1 percent) across the countries in the analysis.

This, however, is only the lowest estimate for the potential contribution of R&D to productivity growth. As the statistically significant and positively signed interaction indicates, the more coordinated the economy, the stronger the impact of R&D in accelerating total factor productivity growth. As Model 6 shows, a one-unit increase in the strength of coordination will allow each 1 percent R&D spending to accelerate productivity growth by 0.47 percent, which is almost half the average acceleration rate across the countries. Even a more modest one standard deviation (0.31) increase on the Hicks–Kenworthy index can add almost 0.14 percent to the acceleration rate of productivity growth.

Figure 3.6 plots the marginal effect of overall R&D spending on the percentage rate of acceleration in total factor productivity growth, conditional on the strength of coordination. As the graph shows clearly, the positive impact of R&D on productivity growth becomes steadily larger as coordination strengthens. When coordination is minimum at zero, 1 percent of GDP increase in R&D spending can accelerate total factor productivity growth by 0.13 percent from last year. As the economy becomes more coordinated, R&D provides

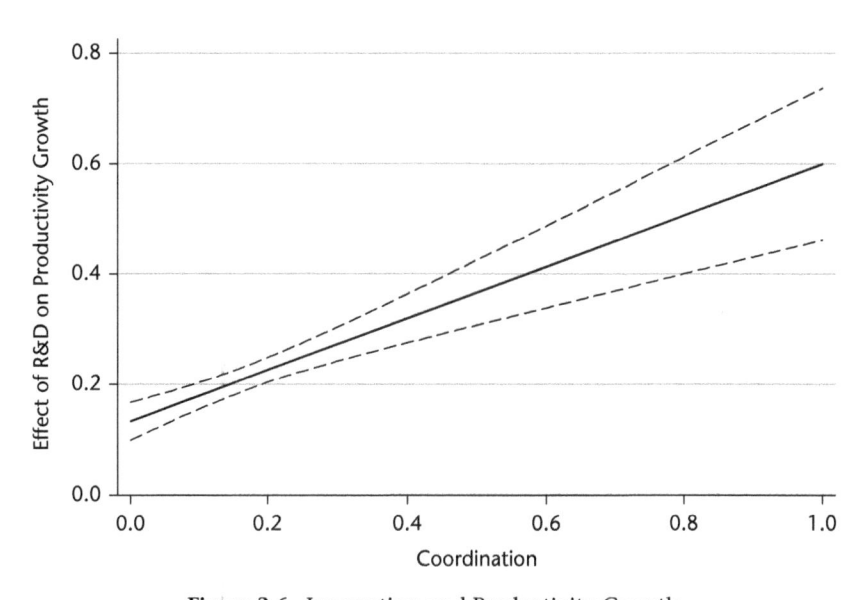

Figure 3.6. Innovation and Productivity Growth

more powerful fuel to the acceleration of productivity growth: at median coordination (close to Italy), each increment of R&D spending can already accelerate productivity growth by 0.34 percent, and by the point of maximum coordination (close to Sweden, Austria, and Norway), the same increment of R&D is sufficient to rev up total productivity growth by 0.6 percent, almost five times the power when coordination is absent. Models 7 through 10 examine the robustness of the findings to various alternative specifications. While Model 6 measured innovation through total R&D spending, Model 7 focuses specifically on applied and experimental R&D, and Model 8 turns to the number of patents. Models 9 and 10 again use Hall–Gingerich to measure coordination in lieu of Hicks–Kenworthy, checking the robustness of findings on R&D (from Model 6) and patents (from Model 8) respectively. Across Models 7 through 10, the interaction between innovation and coordination remains positive, and significant at the 0.01 level. Because these various robustness checks do not significantly alter the findings, I omit separate graphs for them. Overall, this section has provided relatively consistent evidence that strategically coordinated economies are better at using new technologies to boost productivity than liberal market economies. Just as the focus on product innovation in liberal market capitalism pushes the economy on the *extensive* margin (expanding employment), the focus on process innovation in coordinated capitalism pushes the economy on the *intensive* margin (deepening productivity).

Conclusion

New technological knowledge may be used to develop either new products or new production processes. In this chapter, I have argued that different types of capitalism make use of new technological knowledge in different ways. Strategically coordinated capitalism focuses on process innovation, while liberal market capitalism focuses on product innovation. This difference in innovation strategy leads to different outcomes in employment and productivity, two of the most important benchmarks in economic performance. In strategically coordinated economies, innovation raises productivity but has little effect in creating new jobs; in liberal market economies, innovation expands job opportunities but has little effect in lifting productivity. In other words, while innovation is employment-friendly in liberal market capitalism, it is productivity-friendly in strategically coordinated capitalism.

The logic behind why some countries choose product innovation while others choose process innovation lies in the different nature of rents generated by these two types of innovation. In product innovation, new products establish a temporary monopoly in product offering, allowing the innovator to charge a premium for novelty, creating "monopoly rents." In process innovation, new technical knowledge raises productivity (smaller overhead, quicker turnover, and higher output), allowing the innovator to capture "productivity rents" during the production process. The premium for novelty or monopoly disappears when the innovation is shared with other firms. The ability to raise productivity, by contrast, may further increase when firms share process technologies with each other. In other words, process and product innovation thrive in opposite relational environments. While the former generates more rents when firms can commit to deep sharing with relational insiders, the latter is more valuable precisely when firms cannot commit to relational sharing.

By distinguishing between product and process innovation, this chapter has adopted a very different analytical angle on innovation than the commonly discussed typology of radical versus incremental innovation. This new angle helps shed some new lights on the question of economic "dynamism." What type of capitalism is more dynamic? The importance of creating a dynamic economy capable of vigorous growth is not only widely recognized in popular and policy debates, but also an important theme in the comparative political economy literature. For example, much of the literature on neocorporatism and varieties of capitalism aim to better understand the institutional conditions for healthy and dynamic economic growth, not only in aggregate output (such as macroeconomic growth in Hall and Gingerich 2009; Garrett 1998; Franzese 2002), but also in human capital (such as skill formation in Culpepper 2003; Thelen 2004; education in Ansell 2010), employment (Iversen and Wren 1998; Pontusson 2005; Kenworthy 2008), and technological capabilities (Hall and Soskice 2001; Ornston 2012; Casper 2000; Taylor 2004). The importance of "dynamism" in the economy has drawn increasing attention since Penrose's (1995) seminal work on the concept of "dynamic efficiency," which sets a very different criterion for evaluating economic performance than the traditional concern with "allocative (Pareto) efficiency." While Pareto efficiency determines the economy's ability to *allocate existing* resources to their best possible uses, dynamic efficiency reflects an ability to effectively *search for new methods* of generating resources. The "dynamism" of an economy, in other words, is intimately connected to its capacity for innovation.

To the extent that more radical innovation is a reflection of greater "dynamism" (understood in the Penrose framework as the capacity to innovate), the radical/incremental distinction sheds light on the *level* of dynamism in the economy. In other words, by understanding what kind of capitalism engages in "more radical" innovation, we may learn what kind of capitalism is "more dynamic." By contrast, this chapter's product/process innovation distinction sheds light on the *type* of dynamism in the economy. As I discussed earlier, both product and process innovations can be either radical (such as Mac OS X computers and hot-blast methods) or incremental (such as mild hybrid engines and platform-sharing methods in car engineering). Both types of innovation, that is, can be very dynamic or incremental. As a result, the finding that coordinated capitalism has more process innovation does not imply that such economies are necessarily less (or more) dynamic than liberal market economies. The implication, instead, is that liberal market and strategically coordinated capitalism may have *different kinds of dynamism*: the former focuses on the creation of new products while the latter focuses on the development of new production processes. While the dynamism of liberal market capitalism is *extensive* (expanding new product lines and job opportunities), the dynamism of coordinated capitalism is *intensive* (deepening productivity and sophistication in the making of existing products). "Extensive" and "intensive" innovation, in other words, represent two different types of economic dynamism.

The two chapters so far have adopted two distinctive settings to study the political economy of information distribution. Set in *financial markets*, Chapter 2 examined how firms communicate hidden information to investors. Set in *product markets*, Chapter 3 examined how firms make use of their new technological knowledge. Set in *labor markets*, the next chapter studies how firms absorb information about workers' human capital. This, as I will show, have deep implications for one of today's most debated economic outcomes: inequality.

4

Whose Innovation Creates More Inequality?

Today, inequality is one of the most important issues confronting advanced industrialized democracies. Earnings and disposal income inequality are on the rise, albeit to a varying extent, across a larger number of OECD countries (Pontusson 2005; Galbraith 1998; Galbraith and Berner 2001; Alderson and Nielsen 2002). An enormous body of literature has provided evidence from various angles that inequality has potentially far-reaching consequences for the economy and society, including economic growth (Boix 2009, 2010), inflation (Desai et al. 2003), political participation (Bartels 2008; Solt 2008), preference for redistribution (Lupu and Pontusson 2011; Moene and Wallerstine 2003; Kelly and Enns 2010), public goods provision (Baldwin and Huber 2010), and the taste for revolt (MacCulloch 2004; Acemoglu and Robinson 2005; Boix 2003).

This chapter examines the effect of technological innovation on inequality. In particular, it asks: "How does innovation affect earnings inequality differently for different varieties of capitalism?" To address this question, I draw upon another type of inequality which has received somewhat less attention in the comparative political economy literature: unequal attainment in academic education. Scholars have long highlighted how the population's overall level of educational attainment, in both supply and demand, may affect income inequality in the economy. For example, technological innovation is commonly understood to increase the demand for education, and as the supply of educated workers falls (rises) for various exogenous reasons, the wage premium for education rises (falls), increasing (reducing) earnings inequality (Katz and Murphy 1992; Aghion et al. 1999; Acemoglu 2003b). From this perspective, the importance of education is more reflected in the "aggregate level" than the "difference" of educational qualifications across the population. But, in reality, difference in educational qualifications is an important tool of job market competition. Why do job applicants often compete to "stand out" from each other in their academic credentials? And what kind of economies put more pressure on job applicants to "pull ahead of the pack" in academic qualifications? These are some of the key questions to be addressed in this chapter.

Scholars of comparative political economy have certainly not ignored problems of uneven human capital. For example, Rueda's (2007) theory of social democratic insider/outsider divide may be viewed as an account of "skill gaps," separating well-trained and protected "insiders" from low-skilled and vulnerable "outsiders." Similarly, Goldin and Katz (2008) identify growing gaps in educational attainment, in particular an ongoing deterioration in American students' college completion records, as an important factor behind the inequality-increasing impact of technological innovation in the US. Nevertheless, both perspectives focus on inequality driven by *deterioration at the tail* of the human capital distribution: it is harder for everyone to benefit equally from deep training and advanced education when more outsiders are excluded (by labor market rigidities) in continental Europe, and more students in the US are deprived (by poverty and income inequality) of the opportunity to excel in high school or complete college (Berg 2010).

This chapter, by contrast, focuses on an opposite type of educational inequality, driven by *acceleration at the top*. As I will show, some countries drive workers to attain academic education well above the average for their occupation, but others encourage workers to anchor to the average. I argue that, at the extreme upper ceiling of educational attainment (such as master's and doctoral degrees), overeducation is especially prevalent in liberal market capitalism. Because the lack of strategic coordination discourages training, workers lack skills informative of their potential productivity. As a result, they have to use academic credentials to signal productivity, even if they invest in more academic education than will be used in their actual job. While existing studies of human capital inequality place greater emphasis on the problem of "underachievers" (such as high-school dropouts), this book turns its attention to "superstars" in education (such as master's and doctoral degrees). In contrast to existing emphasis on how the tail falls behind the rest, this book sheds some light on how the head pulls ahead of the rest.

This drive towards advanced academic education in liberal market capitalism also affects the course of its technological innovation, increasing radical innovation relative to incremental innovation. On one hand, the newly emerging sectors of radical innovation command a high wage premium; on the other hand, income is redistributed to these sectors in the form of "innovation rents" (i.e. novelty rents from product innovation and productivity rents from process innovation, as noted in the previous chapter). As a result, wealth is redistributed towards the top. This leads to greater inequality of earnings in the economy, by allowing the top to pull ahead of the rest. By contrast, in strategically coordinated capitalism, overeducation is less prevalent, and innovation is less radical. Rent redistribution to sectors of incremental innovation draws median earnings closer to the top, but further away from the bottom of labor-intensive noninnovative sectors.

In a nutshell, liberal market capitalism suffers from two types of *inequality from the top*: an educational gap (top academic attainment well above the average for given occupation), which in turn allows innovation to drive an *earnings* gap (top well above the median in earnings). Because both outcomes pull the top ahead of the rest, I refer to this inequality effect of liberal market capitalism as the *superstar effect*. This forms a contrast to strategically coordinated capitalism, where, as I will show, the bottom falls below the median in earnings as a result of technological innovation (i.e. the *long-tail effect*).

In short, this chapter will examine *two* very distinct forms of "inequality from the top" (one in academic education, the other in wage earnings), and show how technological innovation connects these two outcomes. Although both are examples of "inequality from the top" and both relate to the same process of "radical innovation," they connect to innovation through very different mechanisms. Educational "inequality from the top" reflects the "input" end of innovation: since advanced academic education is a crucial source of human capital for radical innovation, overeducation drives innovation in a more radical direction. Earnings "inequality from the top," on the other hand, reflects the "output" end of innovation: an important byproduct of radical innovation is a widening wage gap between top and median earners in the economy. As a result, although both are connected to radical innovation, educational and earnings inequality are driven by very different independent variables: overeducation is driven by the *lack of strategic coordination* in the economy, while top/median earnings gaps are driven by the *prevalence of radical innovation* in the economy. Meanwhile, these two causal processes are connected by the fact that overeducation makes innovation more radical.

Because of this subtle causal chain, the chapter should not only provide a *distinct* treatment for each of the two forms of "inequality from the top," but also explain how they are connected to a *common* process of radical innovation. Due to the complexity of this task, the chapter also has a more complex structure than earlier chapters. While earlier chapters have only *two* sections, this chapter has *three* key sections: the *first* and *third* sections examine "inequality from the top" at, respectively, the "input" and "output" end of radical innovation, i.e. educational and earnings inequality. A brief *second* section serves as a "mezzanine," examining how one inequality feeds into the other through the process of innovation: overeducation in the workforce drives innovation in a more radical direction. Beyond establishing the linkage between the first and third sections, the mezzanine itself does not directly address the chapter's main theme of capitalism and inequality. For this reason, the mezzanine is also shorter than the other two sections. To help illustrate

```
┌─────────────────────────────────────────────────┐
│              Section (1)                        │
│  Lack of strategic coordination → Overeducation │
└─────────────────────────────────────────────────┘
                       ⇩
┌─────────────────────────────────────────────────┐
│                 Mezzanine                       │
│       Overeducation → Radical innovation        │
└─────────────────────────────────────────────────┘
                       ⇩
┌─────────────────────────────────────────────────┐
│                Section (3)                      │
│        Innovation → Earnings inequality         │
└─────────────────────────────────────────────────┘
```

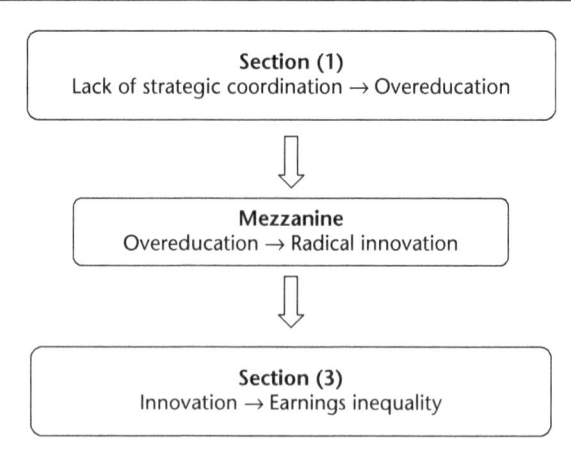

Figure 4.1. The Organization of Chapter 4

the causal chain, in Figure 4.1 I visually outline the causal argument of each section as well as the linkage between sections. As the figure shows, because overeducation is the mechanism that connects patterns of coordination to innovation, it takes a backseat once the chapter's attention turns to innovation's flow-on effect on inequality (section (3)).

Figure 4.1 indicates that the chapter devotes considerable space to developing a causal mechanism (based on overeducation) behind nations' comparative advantage in radical innovation, because radical innovation drives much of the income inequality patterns examined in the chapter. However, was it really necessary to provide a separate theory for radical innovation, given that Chapter 3 already provided another theory of comparative advantage, in product versus process innovation? For example, if advantage in radical (incremental) innovation is merely a manifestation of existing strength in product (process) innovation, then the analytical shift to radical innovation (and the attendant overeducation theory) may have been somewhat gratuitous. However, as Chapter 3 pointed out, the product/process divide cuts across, rather than overlaps with, the radical/incremental divide in classifying innovation. A product innovation may lead to either considerably different products (such as Mac OS X vis-à-vis Windows operating systems) or incrementally better products (such as "mild hybrid" vis-à-vis "full hybrid" car engines). Similarly, a process innovation may be relatively drastic (such as the invention of the "hot blast" technique in iron production and the "float glass" technique in glass production) or gradual (such as the growing use of platform sharing in automobile engineering and design (Gnosh and Morita 2008)). Because the product/process divide is largely orthogonal to the radical/incremental divide, we may have some greater confidence that this chapter's focus on radical/incremental innovation can yield some unique insights on inequality, which are unlikely to be uncovered by solely studying product and process innovation.

By examining inequality, the chapter enters what is now a very crowded field. In the comparative political economy literature, questions of inequality have been examined by a vast number of influential studies from many different angles (Pontusson 2005; Kenworthy 2004; Rueda 2008; Iversen and Stephens 2008; Ansell 2010). A single chapter, of course, cannot address all aspects of this topic. For this reason, before the main arguments unfold, it is important to clarify what the chapter *does not* examine or claim, so that my findings can be transparently situated relative to other studies that adopt very different angles. Four points in particular are worth clarifying at the outset.

First, although this chapter focuses on inequality "from the top" in academic education, it does not in any way claim that inequality "from the bottom" is less extensive. It chooses to focus on top-driven inequality simply because the mechanism for this outcome (job market signaling through educational ranking) is more pertinent to the book's overarching theme about the political economy of information. By contrast, as Goldin and Katz's (2008) influential study pointed out, the mechanism for falling behind at the low end of education (such as transition from high school) is more driven by factors such as regional and income inequality and urban poverty, which fit less naturally into the framework of information distribution on the job market. In other words, although educational inequality "from the bottom" is outside the scope of this chapter, it is by no means a trivial outcome in advanced industrialized economies.

Second, although this chapter examines educational inequality "from the top," it is not a study of higher education (Ansell 2008), because the broad category of postsecondary education encompasses too large a segment of the labor force, only a small portion of which may be meaningfully understood as the extreme upper ceiling, or "top," of the educational ladder. For example, according to the US Census Bureau's (2012) examination of academic educational attainment in 2012, a majority (57.28 percent) of the US population aged 25 and over had at least some college education, and slightly under a majority (40.58 percent) hold an associate's or bachelor's degree. By contrast, a much smaller portion reach more advanced academic education, with 8.05 percent holding a master's degree and 3.07 percent holding a professional/doctoral degree. In fact, as my later empirical analysis based on the International Social Survey Program (ISSP) will show, the average share of population with an educational attainment more than *two standard deviations* above the mean of their occupation (the farthest advance above the mean) is around 13 percent across countries. This figure is not far from the share of the US workforce with a master's degree and above. This *extreme upper ceiling* of educational attainment is what the chapter aims to study with its notion of "overeducation." This is different from the analytical framework of Goldin and Katz (2008), who place greater emphasis on a segment of attainment closer to the average: successful transition from high school to college. As a result, there is no real contradiction between the book's argument of overeducation in liberal market economies (such as the US) and Goldin and Katz's finding that education in the US has fallen *behind* demand. In fact, these two accounts are complementary, bringing together different educational segments to paint a more complete picture of educational inequality in liberal market economies: too many people "fall short" in making a successful transition from high school to college, but those that make it tend to "overshoot," getting academic qualifications well above the typical for their occupation. In other words, inadequate supply (at the tail) and overeducation (at the top) are both present in such economies.

Third, this chapter is a study of inequality in educational *attainment*, rather than *performance*. In other words, the "top" (or for that matter the "bottom") of educational ladder in this chapter is based on how advanced the academic degree is, rather than how effectively the academic knowledge has been absorbed and applied by students. In fact, this chapter's argument about overeducation relies precisely on the fact that academic qualifications and actual performance are *not* perfectly correlated. In other words, when workers engage in job market signaling, they need a higher degree than others even if this does not translate into real increase in productivity on the job. As a result, they become overeducated for their job. For this reason, the chapter's analytical angle is very different from studies that compare good and bad *performers* within a *common* educational segment, such as the OECD International Adult Literacy Survey (IALS), which breaks down performance on "basic" skills (language, math, and document interpretation) into top and bottom percentiles (Iversen and Stephens 2008; Nelson and Stephens 2012). As is well known, IALS data show that while Anglo-Saxon countries are not much ahead of Europe in the performance of the top-fifth percentile, they fall seriously behind in the performance of students at the bottom-fifth percentile. However, there is no real contradiction between this finding and the book's argument about "over-the-top" education in Anglo-Saxon countries, because the two accounts adopt very different notions of "top" in education ("high scores" from basic skill tests in IALS versus "high academic degrees" in this chapter). In fact, by focusing on a literacy level close to the Scholastic Assessment Test (SAT) used in college *entrance* (Nelson and Stephens 2012), IALS covers a segment of educational attainment far removed the analytical focus of this chapter.

Fourth, this chapter is not a comparison of *earnings inequality* across varieties of capitalism. Instead, it is a comparison of *the effect of technological innovation on earnings inequality* across varieties of capitalism. This sets the book apart from many other works on inequality in affluent capitalist economies, where the *level* of earnings inequality is a key outcome of interest (Pontusson 2005; Kenworthy 2004, 2008; Rueda 2008; Kelly 2005; Moene and Wallerstein 2003; Lupu and Pontusson 2011). On the level of inequality, it is well known that liberal market capitalism has greater inequality both from the top and from the bottom than strategically coordinated capitalism. For example, both the top/median earnings gap (p90/p50, i.e. the ratio of earnings for workers at the ninetieth percentile over the earnings for workers at the fiftieth percentile) and the median/bottom earnings gap (p50/p10) are larger in the US than Nordic countries, for every year of available data from the OECD Employment and Labor Market Statistics database. In the book, by contrast, the main outcome of interest is not the *level* of inequality, but the *slope* between inequality and innovation. Instead of asking what countries have a wide p90/p50, so to speak, the book asks what countries allow technological innovation to widen p90/p50. On this, the book argues that innovation has different effects on inequality in different types of capitalism: in liberal market capitalism, new technologies widen earnings inequality from the top (p90/p50), but in strategically coordinated capitalism, they widen inequality from the bottom (p50/p10). Again, there is no real contradiction between this argument and the fact that both p90/p50 and p50/p10 are larger in liberal market capitalism. Instead, they are complementary, bringing together different facets of inequality (level and slope) to paint a more complete picture of inequality in affluent capitalist economies: while Anglo-Saxon countries suffer higher inequality than (especially Nordic) Europe from both ends (top and bottom), new technologies *blunt* one end of this difference (making Europe more unequal) but *sharpen* it at the other end (making Anglo-Saxon countries even more unequal).

Who Are Overeducated?

Between liberal market and strategically coordinated capitalism, whose workers are more overeducated in academic education? While the comparative political economy literature has rarely discussed this question, it has devoted enormous efforts to another question from the opposite end: whose workers are more undertrained in skills (Lynch 1994; Crouch et al. 1999; Culpepper 2003; Thelen 2004; Martin and Swank 2012)? Skills and academic knowledge are complementary forms of human capital: better-educated workers are easier to train, and better-skilled students absorb academic knowledge more effectively (Rosen 1976; Heckman 2000). As a result, it is natural to expect that the development of vocational skills and academic knowledge should move in lockstep with each other: if one falters (or pulls well ahead), the other should follow, too. Based on this logic, do the obstacles that prevent an economy from adequately training its workers also cause it to create academic underperformers at school? Do economies that encourage the training of high skills also drive workers towards the top of the academic educational ladder? This section, however, argues precisely the opposite: despite all the inherent synergies between skills and academic knowledge, training and academic education do not move in parallel. In liberal market capitalism, they move apart: workers' skills *fall short* of the typical demand of their job, but their academic education *exceeds* it. In strategically coordinated economies, by contrast, training and academic education move closer towards each other: workers neither underachieve in skills nor overachieve in education.

How countries compare in the skill training of their workforce has long been an important subject in the comparative political economy literature. On this topic, the relative weakness of training in liberal market Anglo-Saxon economies has been extensively documented by many quantitative and qualitative studies (Iversen 2005; Martin and Swank 2004, 2012; Thelen 2004; Iversen and Soskice 2010). This book does not intend to duplicate these well-documented findings on skill formation. Nevertheless, it is job applicants' lack of skills that gives rise to job market signaling using academic credentials. As a result, in order to fully develop the logic behind my theory of over-education, I start with a brief discussion of theories and evidence on skill formation from the existing literature.

Who Are Undertrained?

Deep skills are very valuable assets for firms as well as workers. However, as both students of comparative political economy (Hall and Soskice 2001; Soskice 1994) and mainstream economics (Becker 1993; Acemoglu and Pischke 1998) point out, the job market on its own may fail to supply skills adequately, because firms face various disincentives for investment in training. For example, firms usually cannot fully capture the benefits from training, especially from skills portable beyond the firm. When skills are portable, they become potential "lootable assets" that all firms are tempted to raid from each other (Soskice 1994). The more portable the skills, the more difficult it is for a firm to fully internalize the benefits of training its own workforce, and the greater its incentives to shirk and loot from others (Streeck 1989; Stevens 1996).

Apart from firm-side obstacles, there are also training disincentives on workers' side. Especially when skills are deep and asset-specific, workers may become trapped inside their own firm by their skills, because they are not able to apply these skills usefully elsewhere. As a result, they face risks both inside and outside their firm. From the inside, the firm can attempt to reduce workers' surplus (be it wages, working conditions, or decision-making influence), exploiting its bargaining power from this situation of "holdup." From the outside, workers exiting their current firm are potentially exposed to sustained unemployment because of their difficulty in finding another job with compatible skills (Iversen 2005; Klein et al. 1996).

In a nutshell, in the absence of institutions that may help overcome these various training disincentives for employers and workers, the economy may fail to invest adequately in skills despite their obvious benefits. In response, a large and growing body of work has focused on possible institutional solutions to this problem of underskilling. Although different studies adopt different analytic perspectives and empirical settings, most concur on the important role of interfirm or firm–labor coordination in boosting the incentives for deep training.

For example, as Martin and Swank (2012) point out, strategically coordinated capitalism, especially its institutions of business coordination based on centralized federation authority and encompassing associational membership, encourages deep training. Martin and Swank identified three important mechanisms through which coordination may facilitate training. First, through a "revenue effect," coordination among businesses, state, and labor delivers wage restraint and raises firms' expectation of what they may gain from highly productive labor. This serves to reduce the extent to which firms discount the future payoffs of training, in effect extending their horizon for human capital investment. Second, through a "social effect," coordination enhances collective action among firms in skill investment, and curbs the incentive to free ride and poach skilled workers from each other. Third, through a cognitive "agenda-setting effect," institutions of coordination (such as business federations or industrial institutes) allow otherwise disparate firms to communicate directly through a common discourse based on long-term skill investment. This feeds back into the "social effect" of promoting collective action, providing a "focal point" for firms to coordinate around.

While theories above center on firms' desire to train, various works by Iversen and associates (Estévez-Abe et al. 2001; Iversen 2005) have focused on workers' parallel desire to get trained. This analytical perspective has shed important light on the role of coordinated capitalism in insuring workers against the potential risk of investing in asset-specific human capital. Two mechanisms (one direct and the other indirect), in particular, allow coordination to play such a role. First, through a "direct effect," business–labor coordination encourages firms to offer skilled workers various benefits that offset both the "outside risks" of sustained unemployment (such as generous social security benefits), and the "inside risks" of holdup by the current firm (such as better wages and better working conditions). Second, through an "indirect effect," the risks of these benefits being opportunistically reversed ex post are mitigated by the presence of parties strongly tied to their core constituencies, as well as governments based on coalition and compromise. Stemming from electoral systems based on proportional representation, these political "background conditions" set coordinated capitalism apart from their liberal market counterpart, which is characterized instead by first-past-the-post electoral systems (Iversen and Soskice 2006, 2009; Martin and Swank 2011). In liberal market Anglo-Saxon countries, parties tend to vie for the median voter, and governments tend to comprise of single parties. As a result, drastic policy reversal is more likely in such countries, which reduces the credibility of policies aimed at insuring workers against risks associated with deep skill investment.

While the comparative political economy literature focuses mainly on the inadequate *level* of skill investment when coordination is insufficient, students of labor economics point out that the very *portability* of skills may also be affected. Anticipating other firms to raid one's own skilled workers, a firm will prefer to invest only in firm-specific skills, which become obsolete once outside the firm, and hence make the firm's own workers unattractive targets for poaching (Müller and Shavit 1998; Stevens 1996). There are three important implications from this additional perspective on portability. First, low portability may lead to even lower level of skills. As firms skew their training further towards firm-specific skills, they increase further the "inside risks" of holdup and "outside risks" of sustained unemployment for their workers (Iversen 2005; Klein et al. 1996), which further erodes workers' incentives to train. Second, both low portability and low level of skills may intensify the use of academic qualifications for job market signaling. When a job applicant carries few skills or only skills that expire beyond the previous firm, these skills are not informative indicators of her potential productivity for other firms on the job market. This increases the need for employers and job applicants to rely on academic credentials as productivity signals (Thurow 1976; Borghans and de Grip 2000b). In other words, low portability of skills further reinforces the chapter's expectation of overeducation in liberal market capitalism.

Third, labor economists' suggestion that weak coordination drives firms to invest only in skills that expire beyond the firm apparently jars against arguments from the comparative political economy literature that human capital is *more* portable in liberal market economies. For example, in a well-known passage from Hall and Soskice's (2001, p. 17) foundational work on varieties of capitalism, they suggest that liberal market economies encourage workers to invest in "switchable assets," while coordinated economies encourage "specific or co-specific" human capital assets. However, the tension between these two accounts is more apparent than real, because a careful reading of the comparative political economy literature shows that it openly embraces the fact that skills are more encompassing in strategically coordinated economies (Iversen and Stephens 2008). For example, Germany, as a quintessential representation of coordinated capitalism, has a dual apprenticeship system which makes substantial investment in standardized (i.e. industry-specific) rather than firm-specific training (Herrigel and Sabel 1999; Culpepper 2003; Finegold and Wagner 1999).

For this reason, the comparative school's argument that liberal market capitalism has more mobile human assets should not be interpreted as a narrow comparison of "skills." Instead, it is a broad comparison of "human capital," taking into consideration the weight of both skills and academic knowledge. Because workers in liberal market economies have inadequate skills (either because they are undertrained or because they are trained with skills that become obsolete beyond the previous firm), they rely more heavily on academic credentials to signal productivity. Although vocational skills may be broad, their scope is delimited by the specific demands of the industry in which they are applied. Theoretical knowledge from natural sciences (such as math and physics), by contrast, may be usefully applied in combination with skills from vastly disparate industries and occupations. Academic knowledge, in other words, is a more portable form of human capital than skills acquired through training (Müller and Shavit 1998). As a result, although workers in liberal market capitalism may carry narrow (firm-specific) skills, their human capital portfolio as a whole may still be very portable, because they put greater weight on generic academic education.

However, given the complementarity between skills and knowledge (Rosen 1976; Heckman 2000), how far can workers in Anglo-Saxon economies really advance in academic education when they are poorly equipped with skills? Do poor skills also forestall progress in academic education? As I will show next, instead of dragging educational attainment down, poor skills surprisingly result in *overachievement* in academic education: workers end up with *more* advanced academic education than typically used in their occupation. The labor force, in other words, becomes overeducated. Next, I develop a theory of why some countries have more overeducated workers than others. Then, I take this theory to test, using data from ISSP to demonstrate that overeducation is more pervasive in liberal market economies.

A Theory of Overeducation

As Borghans and de Grip (2000a) review in detail, there is a growing literature exploring the causes and consequences of overeducation, motivated by the influential "job queue" theory of Thurow (1976) and "signaling" theories of Spence (1973) and Akerlof (1976). As this literature points out, overeducation arises when workers use the relative ranking of academic qualifications to signal their otherwise unobservable productivity. As Chapter 2 suggested in the setting of financial markets, the potential quality of innovation projects often cannot be unobserved by investors. In this chapter's setting of job markets, employers face a similar problem of being unable to observe job applicants' potential productivity. Employers naturally want to hire more productive workers, but the potential productivity of job applicants is hidden information for them. How do they separate the future "lemons" (unproductive workers) from productive workers on the job market? Similarly, how do good potential workers avoid being "pooled together" with bad ones on the job market? Because workers' proficiency in production grows with their training, skills become a key source of information that guides the matching process between firms and workers. Workers compete based on training; better-trained workers are in greater demand (Akerlof 1970, 1981).

However, if job applicants bring to the labor market little background of training, or bring only skills that have already expired (beyond the previous firm), employers will face difficulty in determining the real productivity of these potential workers. In order to credibly reveal their own productivity and avoid being pooled together with those who are less productive than themselves, job applicants invest in academic educational credentials as a costly "signal" (Spence 1973). Because less proficient or capable individuals are less likely to advance in academic attainment, higher academic degrees become a reasonably informative signal about the applicant's productivity, too costly for the unproductive to mimic. Of course, Michael Spence's seminal signaling thesis has found many empirical applications beyond the labor market. In Chapter 2, for example, I have shown that firms planning to initiate long-term deep relationships with investors send "insider information" as a signal of commitment. Similarly, in wage negotiations, unions initiate periodic conflicts in order to signal their bargaining power (Hall and Thelen 2009; Lundvall 2013). In this chapter, I focus specifically on signaling in the labor market, the original setting for Spence's theory, because in this environment marked by competition (between workers for good jobs), signaling has particularly interesting consequences. Here, although signaling performs the traditional role of "sorting" (good jobs to good workers), it also drives competition for good jobs in an unexpected, bad, direction, turning it into wasteful *rat races*.

With a positive level of unemployment (and therefore a shortage of jobs relative to applicants), job applicants have to compete with each other: higher-ranked applicants get better spots, and with fewer spots than applicants, the bottom ranked will be left without a job (Akerlof 1981; Bowles and Gintis 1990). Competition for jobs is good for employers: it encourages job applicants to invest in higher productivity in order to beat out competitors. However, when applicants use academic credentials to signal their productivity, the competition for jobs starts to acquire the features of a rat race. This occurs because academic educational attainment is a noisy signal for productivity. With some exceptions, most academic knowledge cannot be applied directly to the production process without being reconfigured into (or combined with) skills (Müller and Shavit 1998). Consequently, the *content* of workers' academic education cannot be mapped informatively onto their potential productivity. On the other hand, better-educated workers are more likely to be cognitively capable and cost less to train, so higher *ranking* in academic attainment is an informative signal about higher potential productivity. As a result, firms infer job applicants' potential productivity based on their *relative position* in the educational queue rather than the *content* of their academic education (Spence 1973; Thurow 1976; Borghans and de Grip 2000b).

Because the relative ranking of their academic qualifications sends a signal about their productivity, workers face the constant pressure to rank above others, even if increases in formal education are insufficient to lift their actual productivity. Job applicants know that, even if their real productivity is unchanged, if they present a higher academic degree they will be *pooled* in the job market together with applicants whose productivity is *above* them. Conversely, even if their real productivity does not decline, if they have a lower academic degree they will be *pooled* together with those whose productivity is *below* them. As a result, all applicants face strong pressure to obtain higher degrees (Akerlof 1976). The less information employers can glean from job applicants' existing skills, the more the latter have to rely on the relative ranking of academic qualifications to signal productivity, even if higher degrees do not necessarily translate into higher productivity. The competition for jobs, in other words, degenerates into rat races: workers have to run ever faster (by investing in ever-higher academic education), but the treadmill (productivity on the actual job) remains stationary.

The same logic of rat race also explains another well-known puzzle: why do workers on an assembly line often move with very fast speed (Nye 2013)? While the employer cannot effectively observe workers' actual productivity, she can more accurately observe their *speed relative to each other* (since they are all on the same assembly line). Since faster workers are likely to be more productive, workers are willing to move faster, even if hastier moves may in reality reduce quality and output. By moving with faster observable speed, workers are able to *pool* with those whose real productivity is *above* them; by slowing down, they become *pooled* with those whose real productivity is *below* them (Akerlof 1976). In both the education example and the speed example, workers move up in observables that are only imperfect correlates of unobservable productivity, which remains unchanged in reality.

While this overinvestment in education is driven by the noise in the *ranking* of academic qualifications as productivity signals, it is further exacerbated by the noise in the *content* of academic qualifications as productivity signals. As noted earlier, because academic knowledge often cannot be applied directly to the production process, employers cannot map academic education directly onto future productivity, although they expect better-educated job applicants to be more productive (Borghans and de Grip 2000b). In other words, potential employers expect some (imperfect) positive *slope* between academic education and productivity, but without being able to match specific education to specific productivity, employers cannot identify the *intercept* that anchors this slope. As a result, applicants may be *sorted* on the job market (more educated workers end up with better jobs), but jobs and applicants are not *matched* (workers are more educated than their jobs need).

In short, when the lack of skills informative of potential productivity drives workers to use academic qualifications as productivity signals, they may end up attaining higher academic qualifications than needed by their occupation for two reasons. First, higher *ranking* of educational attainment allows low-productivity job applicants to "compete like a high-productivity applicant," pooling with stronger competitors on the job market. Second, the *content* of academic education does not give employers sufficient information to match specific educational qualifications with specific productivity. For these two reasons, the more the job market depends on academic signaling, the more likely is overeducation.

As noted earlier, the comparative political economy literature has shown extensively that workers in liberal market capitalism receive less training. Along the same vein, labor economists have argued that when these workers do receive training, their skills tend to expire beyond the previous firm (Stevens 1996). Because these job applicants carry little skills informative of their unobservable productivity, they rely heavily on academic education as productivity signals, and as a result overeducation is more likely. By contrast, in strategically coordinated capitalism, firms are willing to train more deeply, and provide skills encompassing enough (i.e. industry-based) to be useful beyond the specific firm (Herrigel and Sabel 1999; Iversen and Stephens 2008). As a result, information about job applicants' productivity is less opaque on the job market. This reduces their need to invest in academic degrees as signals, which in turn reduces the likelihood of overeducation. In other words, the more coordinated the economy, the less likely is overeducation.

> *H1: the more coordinated the economy, the less likely is overeducation in the workforce.*

This chapter argued that the intensity of job market signaling (and behind this, the lack of skill training) is the key explanation for why countries differ in the extent to which their workforce is overeducated for their jobs. However, could there be alternative explanations for the outcome predicted in hypothesis H1? After all, the literature on varieties of capitalism already points out that Anglo-Saxon economies place greater emphasis on general, as opposed to asset-specific, skills. This institutional feature in itself would have led to greater employer *demand* for university education, independent of the signaling mechanism proposed in this chapter. In a similar vein, the intense educational race in the US is also partly driven by institutional characteristics that affect incentives on the *supply* side: de-industrialization, exposure to new social risks, inequality of opportunities (McCall 2013; Armingeon and Bonoli 2006), all of which may encourage workers to hedge against career risks by "loading up" on more academic credentials. As a result, it is important to clarify how the chapter's signaling-based theory differs from these existing theories of university education.

While the above theories of high demand and supply for education predict high levels of educational attainment in the US, the chapter's signaling-based theory makes a more subtle prediction: educational attainment *higher than the industry's own typical need*, which more directly encapsulates the notion of overeducation (a main theme of this chapter). While *high attainment* in education may be explained by high demand and supply, *relative excess* (higher than the industry's own typical need) requires an explanation that incorporates a mechanism of *relative comparison*: when skills are absent, the job market relies on the *relative ranking* of academic attainment to infer workers' potential productivity. In response, workers invest in ever-higher academic degrees, even if the degree exceeds the actual needs of the job. In other words, the signaling theory has greater leverage in explaining the subtle outcome of overeducation than existing theories that focus on aggregate levels of demand/supply for education.

Through overeducation, liberal market capitalism creates a distinct type of educational inequality, pulling the top well ahead of the occupation's own average in academic attainment. For this reason, I refer to such effect of liberal market capitalism as the *superstar effect*. As hypothesis H1 points out, the more coordinated the economy, the less overeducation, and hence the smaller the *superstar effect*.

Before I take H1 to empirical test, it is important to emphasize that, even within the same country, the likelihood of overeducation may vary across different occupations because of their different standards for human capital. An advanced academic degree in natural sciences, for example, may be above normal for routine production, but may be right on the equilibrium for tasks of R&D. The higher the human capital requirement of the occupation, the less room for academic attainment to exceed the occupation's needs, and hence the less likely is overeducation. As a result, in later empirical analyses of overeducation, it is important to control for the general skill level of given occupation. Another implication from this "ceiling effect" on overeducation is that "excessive" academic education is especially unlikely in sectors of radical technological innovation, which relies heavily on advanced research in basic sciences (Griliches 2000; Mansfield 1995; Helpman 1998). Advanced academic education, in other words, is the norm, rather than exception, for radical innovation. The importance of advanced academic education to radical innovation has direct implications for what countries will do better in radical innovation. In liberal market economies, the race towards top education may have resulted in the underutilization of human capital (i.e. overeducation) for routine production or incremental innovation, but the larger number of education high-achievers produced across the economy should also enable such countries to pull ahead in radical innovation. This, in turn, will have direct consequences for earnings inequality. I defer the discussion about innovation and inequality to later. Here, I focus instead on the empirical evidence for overeducation, and show that this phenomenon is indeed more prevalent in liberal market economies.

Empirical Analysis

In the overeducation literature there are three key methodological approaches to measuring overeducation: expert evaluation by professional job analysts (Wolff 2000; Rumberger 1981), self-assessment by workers (Green et al. 2000), and identifying the "upper tail" in the distribution of educational qualifications for given occupations (Verdugo and Verdugo 1989; Battu and Sloane 2000). Although each of the first two approaches has its own distinctive advantages, their practical application has been limited to small-scale studies based on individual industries or even specific companies, because of the high demand on the content of information supplied by the subjects of study. In order to compare the extent of overeducation systematically across a wide range of OECD countries, I follow the third approach based on the distribution of academic qualifications across individuals. This distribution-based approach allows me to tap into the rich set of individual-level data on education and occupation that can be gleaned from ISSP for a large number of countries and years.

For each wave, ISSP consistently taps respondents' detailed occupational category (International Standard Classification of Occupations (ISCO) 88 four-digit) and level of academic educational attainment. Although the data covers a large number of countries, whether a person's academic attainment is "typical" or "above average" should be evaluated on the basis of her own country rather than all countries in the data set, because even for the same occupation, the typical level of human capital used (and hence the benchmark for assessing overeducation) will differ across countries. Although ISSP's occupational data can be disaggregated all the way to the fourth digit under

ISCO, in practice the most feasible level of disaggregation for calculating the distribution of academic qualifications is on the two-digit level. On the three-digit level, the number of respondents is already down to fewer than ten for some occupations, which leaves too few observations to calculate an informative distribution of academic attainment for a given occupation.

Each wave of ISSP has two variables that tap the academic educational attainment of respondents. Of the two measures of academic attainment, the first is based on a relatively simple, cross-country, coding procedure that collapses each country's various levels of academic attainment into one common index with four ranked categories (1 through 4). The other, by contrast, is based on country-specific coding, which preserves the various original academic degrees in each country, and as a result produces a much more refined ranking scale, with many more categories available. By forcing all countries into a common four-category scale, the first (cross-country) measure cannot capture any country-specific feature of academic qualifications. Furthermore, the crudeness of the four categories also makes it difficult to calculate a meaningful distribution: for many occupations, almost all observations for academic attainment fall into only two categories. As a result, I use the second (country-specific) measure of academic attainment, which is also appropriate because whether a person's academic education is typical or above average should be evaluated on the basis of her own country.

Separately for each country, survey wave, and occupation (ISCO two-digit), I code respondents whose academic educational attainment is *two* standard deviations or more above the mean as "overeducated." Theoretically, of course, I can use an even higher bar (three standard deviations or more), but in reality no two-digit occupation in the data has respondents three standard deviations above in academic attainment. In other words, two standard deviations capture the farthest advance beyond average in academic attainment for each occupation. According to hypothesis H1, the more coordinated the economy, the lower a respondent's likelihood of being overeducated. Coordination, in other words, continues to be the key independent variable of interest. Consistent with previous chapters, I continue to use two different measures of coordination (Hicks–Kenworthy and Hall–Gingerich) to increase the robustness of findings.

Besides, I also include a series of additional controls in estimating the likelihood of overeducation. I start by controlling for the age of the respondent. Younger workers may be more susceptible to overeducation in academic qualifications than their older counterparts. First, because they have a longer horizon for the returns of human capital investment (Oster et al. 2013), they may be more willing to spend the longer time and greater resources needed for attaining more advanced academic degrees. Second, relative to older workers, younger workers may be better placed to advance further in academic education because of their more recent experience at school, greater cognitive ability, and greater capacity to absorb new theoretical knowledge. I also control for gender. Compared with men, women are more likely to find employment in nonproduction service-oriented industries, which are relatively polarized in the distribution of skills and pay (Sainsbury 1999; Gornick and Meyers 2003; Wren 2012; Iversen and Rosenbluth 2010; Estévez-Abe 2006). At the "high quality" end are professions such as R&D, finance, law, teaching, and healthcare, all of which rely heavily on advanced academic knowledge, and as a result reduce the possibility for overeducation. At the "low quality" end are sectors such as retailing, catering, home care, or personal services, where productivity and skill input are not only low but also difficult to improve. As a result, employers from this segment of the job market care more about lowering the cost of labor than raising its productivity (Esping-Andersen 1999). For example, a company would rather pay less to hire a high-school graduate as a janitor than pay more for a college graduate to do the same low-end job. This again reduces the need for job applicants to signal higher productivity through higher academic degrees.

More generally, the likelihood of overeducation in an occupation is always contingent on this occupation's own level of requirement for skill and productivity. The higher the human capital requirement of the occupation, the less room for academic attainment to exceed the occupation's needs, and hence the less likely is overeducation. Therefore, it is important to control for the general level of skill for different occupations. The International Labor Organization (ILO) ranks the ISCO occupations into four broad levels of skills, each corresponding to a typical level of academic education requisite for that skill level, based on the United Nations Educational, Scientific, and Cultural Organization (UNESCO)'s International Standard Classification of Education (ISCED). The first broad skill level corresponds to ISCED category 1, covering primary education beginning at the age of five and up to the age of twelve; the second skill level corresponds to ISCED categories 2 and 3, covering first and second stages of secondary education beginning at the age of eleven and up to the age of eighteen; the third skill level corresponds to ISCED category 5 (category 4 was intentionally excluded by ILO's classification scheme because it describes vocational rather than academic education), which begins at the age of seventeen and lasts about four years, leading to "an award not equivalent to a first university degree"; the fourth skill level corresponds to ISCED categories 6 and 7, covering education that again begins at the age of seventeen but which "leads to a university or postgraduate university degree, or the equivalent." This measure of skill level controls for the potential "ceiling effect" on overeducation: the more skilled the occupation, the less room for academic attainment to exceed its needs, and hence the less likely is overeducation.

The school-to-work transition literature further suggests that characteristics of both academic and vocational educational systems may affect the possibility for overeducation in academic qualifications (Kerckhoff 1995; Müller and Karle 1993). Three aspects are especially important: (1) whether vocational education is *occupation-specific* (Müller and Shavit 1998); (2) the level of *standardization* in academic education, defined as the degree to which the quality of education meets the same standards nationwide (greater standardization makes the *content* of academic qualifications more informative signals of productivity, which reduces the need to rely on the *ranking* of qualifications as signals); and (3) the level of *stratification* in education, defined as the streaming of students into distinctive tracks towards tertiary versus vocational education, at the early (secondary education) stage (Allmendinger 1989). It is easiest for employers to map jobs onto the *content* of academic qualifications when the country's educational system is *stratified, standardized,* as well as *occupation-specific* as defined above. This reduces the need for job applicants to use the ranking of academic attainment as productivity signals, which in turn reduces the likelihood of becoming overeducated for the job.

Country-level data for these three features of the educational system (occupation-specificity, standardization, and stratification) is available from the collaborative project organized by Shavit and Müller (1998), where country experts across thirteen countries scored their own country based on binary (zero, one) coding for standardization and three-category (zero, one, two) coding for stratification and occupation-specificity. Table 4.1 lists country scores on these three indicators, for all countries included in the empirical analysis.

As Table 4.1 shows, several countries in the ISSP analysis (italicized in the table) are not covered by the Shavit–Müller data set. For these countries, I code their educational system characteristics based on countries within the Shavit–Müller data set that have similar educational systems. It is important to emphasize that the typology of educational systems cannot be mapped onto well-known typologies of welfare states (Scandinavian, continental European, and Anglo-Saxon (Esping-Andersen 1990)) or varieties of capitalism (coordinated and liberal market economies (Hall and Soskice 2001)). Instead, the classification of countries in educational systems exhibits its own distinctive pattern, and even countries that are historically, geographically, or institutionally close to each other do not necessarily belong to the same type of educational systems. The literature on the typology of educational systems suggests the following groupings: Austria and Denmark belong with Germany, Finland with France, Canada with the US, New Zealand with Australia, Norway with Sweden, and finally Portugal and Spain are similar to Italy. To keep the analysis focused, I do not digress into the related literature underlining such classifications, other than referring readers to some important studies

Table 4.1. Characteristics of Education Systems

	Occupation-Specificity	Stratification	Standardization
Australia	1	0	0
Austria	2	2	1
Canada	0	0	0
Denmark	2	2	1
Finland	1	1	1
France	1	1	1
Germany	2	2	1
Ireland	0	0	1
Italy	1	1	1
Japan	0	0	1
Netherlands	2	2	1
New Zealand	1	0	0
Norway	1	0	1
Sweden	1	0	1
Switzerland	2	2	1
US	0	0	0
UK	1	0	0

Nonitalicized countries are directly covered in the Müller and Shavit (1998) data set.
Italicized countries are not covered in the Müller–Shavit data set. They are instead coded on the basis of countries within the Müller–Shavit data set with similar educational systems.

lending support to this typology (Müller and Gangl 2003; Hannan et al. 1996; Allmendinger 1989; Abbott 2000). Because occupation-specificity and stratification are discrete variables with three categories ("low," "moderate," "high"), I create separate dummies for "moderate" and "high" specificity and stratification, leaving the "low" category as the reference category. Because educational standardization is a binary variable ("low" and "high"), I create a dummy for high standardization, with low standardization as the reference category.

Although ISSP is repeated annually, because it is not a panel survey, each respondent appears only once (i.e. in the survey year), rather than being observed across multiple years. In other words, ISSP does not track the variation of any observation over time. As a result, I estimate each survey as a single cross-sectional analysis and use the findings from different survey years as robustness checks against each other, rather than pooling all surveys to produce (spurious) time series data for one single analysis. While ISSP waves after 2000 consistently provide observations for academic attainment on the ISCO two-digit level for somewhere between ten to fifteen countries, the number of such countries drops sharply to only six for the 1999 survey. Because of this sharp break in sample size, I choose the 2000 ISSP as the first wave to be included in the analysis. I choose the 2006 ISSP as the last wave to be included, because this corresponds to the last year for which there are available data for later analyses of innovation and earnings inequality. In Table 4.2, I report logit estimations for the probability of a worker being overeducated (i.e. attaining academic educational qualifications two standard deviations or more above the average for his/her given occupation). In order to more intuitively capture the effect on the outcome, I report the *percentage change* in the odds of overeducation instead of raw coefficients. All analyses in Table 4.2 use the Hicks–Kenworthy index as the measure of coordination.

According to hypothesis H1, the more coordinated the economy, the smaller the likelihood of overeducation. This expectation is borne out in Table 4.2. I start the discussion with the 2000 wave of ISSP (the first column of the table). Here, a one-unit increase in the strength of coordination will reduce the odds of a worker becoming overeducated by 35.12 percent. In other words, the likelihood of overeducation will be slashed by more than a third. This large effect, however, should be put into a proper context: one unit of movement on the Hicks–Kenworthy coordination index encompasses its entire range, from minimum coordination (zero) to maximum (one). A more realistic increment of one standard deviation increase (0.32 units on the index) in coordination will slash the odds of overeducation by 11 percent, still a substantively noticeable effect. Compared with minimum coordination, even an economy with average strength in coordination (0.51 on the Hicks–Kenworthy index) can reduce the odds of its workers becoming overeducated by 17.5 percent.

With all continuous control variables at their mean and dummy control variables at their reference category, a worker in a country with average coordination faces the odds of about one-to-five for becoming overeducated. A standard deviation more coordination will reduce the odds of overeducation to less than one-to-six, and two standard deviations will drive the odds down close to one-to-eight. Moving in the other direction, by contrast, two standard deviations below average coordination will raise the odds of overeducation close to one-to-three. For some further substantive interpretation of the results, I calculate and compare the likelihoods of overeducation specific to different countries, ceteris paribus. For example, the odds of a worker in strongly coordinated Germany (0.80 on the Hicks–Kenworthy coordination index) becoming overeducated for his/her occupation are 26 percent smaller than his/her counterpart in weakly coordinated Canada (0.04 on the coordination index). On the other hand, the odds of overeducation for workers in Australia (0.17 on coordination) are larger than their Dutch (0.58 on coordination) colleagues by 14.5 percent. Similarly, weak coordination in the UK (0.10 on the Hicks–Kenworthy index) raises the odds of overeducation for British workers by around 23 percent above their counterparts in strongly coordinated Japan (0.77 on Hicks–Kenworthy). On the other hand, the risk of overeducation for workers in Austria (0.96 on the coordination index) is 34 percent smaller than the risk faced by US workers (0.01 on coordination) for the same occupation. In short, overeducation in academic attainment is indeed less likely in more strategically coordinated capitalism.

Table 4.2. Percentage Change in the Odds of Overeducation, ISSP 2000–6

	2000	2001	2002	2003	2004	2005	2006
Coordination (Hicks-Kenworthy)	−35.12(0.11)***	−46.03(0.10)**	−29.04(0.03)***	−32.53(0.09)***	−44.35(0.06)***	−24.15(0.05)***	−19.46(0.02)***
High specificity of vocational education	−15.21(0.19)	−24.90(0.14)	−31.25(0.28)	−29.42(0.13)*	−21.79(0.12)*	−34.51(0.07)***	−9.82(0.11)
Moderate specificity of vocational education	−10.16(0.05)**	−27.51(0.08)**	−36.18(0.05)***	−10.01(0.08)	−39.04(0.12)**	−26.30(0.08)***	−11.25(0.09)
High stratification of education	−21.40(0.08)***	−19.53(0.09)***	−35.0(0.08)***	−23.14(0.09)**	−15.36(0.10)**	−23.05(0.03)***	−29.15(0.07)***
Moderate stratification of education	−18.73(0.05)***	†	−21.03(0.06)**	−22.52(10)	−19.27(0.09)**	−15.63(0.11)**	−21.41(0.03)***
High standardization of academic education	19.21(0.21)	25.26(0.29)	24.19(0.36)	15.01(0.27)	12.01(0.19)	33.40(0.28)	16.04(0.38)
Skill level of occupation	28.42(0.02)***	20.06(0.03)***	32.95(0.02)***	30.00(0.03)***	20.23(0.05)***	29.89(0.02)***	17.27(0.05)***
Age	−3.19(0.00)***	−2.54(0.00)***	−2.75(0.00)***	−3.69(0.00)***	−3.04(0.00)***	−2.63(0.00)***	−3.33(0.00)***
Male	15.69(0.03)***	11.51(0.10)***	28.67(0.06)***	25.61(0.03)***	33.16(0.07)***	29.35(0.04)***	34.25(0.06)***
P-value (likelihood ratio test)	0.00	0.00	0.00	0.00	0.00	0.00	0.00
Number of Countries	12	10	15	15	15	15	15
N	10,207	8959	12,753	13,192	14,259	14,823	15,898

* p<0.1; ** p<0.05; *** p<0.01

† Dropped due to perfect collinearity with occupation-specificity of vocational education.

This pattern from the 2000 ISSP is corroborated by similar findings from the other six waves of ISSP spanning between 2001 and 2006. Glancing at the first row in Table 4.2 across all seven surveys, the percentage change in the odds of overeducation is consistently negative, and significant at either the 0.01 level (in six surveys) or the 0.05 level (in one survey). Depending on the year of analysis, an increase from minimum to maximum coordination can slash the risk of overeducation by somewhere between a fifth (19.46 percent) and almost half (46.03 percent). In other words, each one standard deviation increase in coordination can reduce a worker's odds of overeducation by somewhere between 6.21 percent and 14.73 percent. Again, holding all continuous control variables at the mean and dummy control variables at reference category, the odds of overeducation are somewhere between one-to-five and one-to-four (depending on the year of survey) for a worker in an economy with average strength of coordination. A standard deviation more coordination will reduce the odds to somewhere between one-to-six and one-to-seven, and two standard deviations will drive the odds down to as low as one-to-nine.

In order to establish the robustness of findings, Table 4.3 repeats the analyses from Table 4.2 on the same set of data, but using the Hall–Gingerich index instead of Hicks–Kenworthy to measure the strength of coordination. As Table 4.3 shows, the core findings are not qualitatively altered. Depending on the year of the survey, moving from minimum to maximum coordination (as measured by the Hall–Gingerich index) can reduce a worker's odds of being overeducated for his/her occupation by somewhere between 19.38 percent and 42.86 percent. In other words, each standard deviation increase in coordination can reduce the likelihood of overeducation by somewhere between 6.2 percent and 13.72 percent. Relative to average strength in coordination, a liberal market economy with minimum strategic coordination will increase the likelihood of overeducation by somewhere between 9.88 percent and 21.85 percent.

This finding of overeducation in liberal market economies can be further reinforced by evidence from the case study literature, which helps shed some indirect light on the micro-level causal mechanism behind this outcome. To the extent that the nature of student and employer demand for education shapes universities' strategies in supplying education, Lukas Graf's (2008) focal comparison of two universities (University of East Anglia in the UK and University of Kassel in Germany) in their strategies to break into the international market illuminates the different types of incentives faced by students (as well as industry) in these two countries. As the Pro-Vice Chancellor of East Anglia pointed out, the university's competitive strategy on the global market is heavily influenced by patterns of domestic competition in the UK's own job market. According to this chapter's argument, job market

Table 4.3. Percentage Change in the Odds Ratio of Overeducation, Additional Analysis

	2000	2001	2002	2003	2004	2005	2006
Coordination (Hall-Gingerich)	−22.13(0.05)***	−34.92(0.08)***	−42.86(0.05)***	−19.38(0.04)***	−40.38(0.03)***	−37.21(0.06)***	−31.05(0.02)***
High specificity of vocational education	−21.47(0.01)***	−29.50(0.05)***	−20.00(0.24)	−26.51(0.16)	−34.17 (0.10)**	−27.58(0.12)*	−12.27(0.15)
Moderate specificity of vocational education	−28.65(0.12)*	−20.02(0.03)***	−21.49(0.03)***	−36.17(0.11)	−17.98(0.06)**	−24.81(0.05)***	−13.62(0.08)
High stratification of education	−12.31(0.02)***	−27.48(0.05)***	−25.36(0.06)***	−26.53(0.09)**	−34.53(0.03)***	−20.38(0.06)***	−15.93(0.04)***
Moderate stratification of education	−20.51(0.14)	†	−25.82(0.04)*	−12.40(12)*	−22.70(0.13)*	−36.53(0.15)	−27.50(0.01)***
High standardization of academic education	11.01(0.25)	39.64(0.36)	21.53(0.41)	19.96(0.31)	7.64(0.23)	20.68(0.35)	13.52(0.26)
Skill level of occupation	21.83(0.05)***	18.62(0.01)***	25.47(0.03)***	31.04(0.03)***	14.41(0.04)***	−28.36(0.02)***	−24.16(0.04)***
Age	−2.92(0.00)***	−2.64(0.00)***	−2.39(0.00)***	−3.68(0.00)***	−2.81(0.00)***	−2.20(0.00)***	−3.53(0.00)***
Male	29.05(0.08)**	36.57(0.11)***	20.02(0.05)***	20.97(0.02)***	16.80(0.03)**	24.16(0.02)***	34.13(0.04)***
P-value (Likelihood ratio test)	0.00	0.00	0.00	0.00	0.00	0.00	0.00
Number of Countries	12	10	15	15	15	15	15
N	10,207	8959	12,753	13,192	14,259	14,823	15,898

* p<0.1; ** p<0.05; *** p<0.01
† Dropped due to perfect collinearity with occupation-specificity of vocational education.

competition in liberal market economies relies strongly on "credential rankings" in academic education, which are the micro mechanism driving the outcome of overeducation. East Anglia's internationalization strategy provides a vivid example for this pattern of educational competition in the UK. Although human capital is a multidimensional asset whose value varies by how well it matches with the job, East Anglia compresses the assessment of educational quality into quantitative one-dimensional indicators that can be scored entirely separately from industry and occupational settings. In particular, the university pays strong attention to "league tables and rankings," in a manner similar to the public monitoring of companies using a "limited number of quantifiable variables and standardized 'benchmarks', and ranked according to weighted share performance" (p. 50). The University of Kassel in Germany, by contrast, adopts a different approach to competition on the educational market. Instead of trying to "outrank" other institutions in a vertical contest, Kassel seeks to deepen lateral collaboration with other institutions as associates. While relative "credential rankings" are less emphasized, German universities pay stronger attention to the "match" between education and the actual requirements of the job. The various "foreign satellites" of German universities are good examples of the tight connection between academic education and industry needs. The Thai–German Graduate School of Engineering in Bangkok, for example, was established by RWTH Aachen University. Since Thailand is an important production site for German automobile manufacturers, various important players in the industry are involved with the graduate school, including Siemens, ABB, Bayer, BMW, and Mercedes-Benz. Similarly, the German Institute of Science and Technology in Singapore was established by Technical University Munich. Because Singapore has a strong pharmaceutical industry, the institute focuses on offering industrial courses, and receives funding, traineeships, research chair positions, and research contracts from various German companies in the chemical sector, including Degussa, Wacker, Merck, BASF, Bayer, Celanese, Südchemie, and Allianz.

The contrast between the emphasis on formal education in liberal market economies and industry-specific skills in coordinated economies can also be found within country, on a subnational level. As noted earlier in Chapter 3, Asheim and Iskasen (2002) examined Norway's regional variation in patterns of innovation, highlighting a contrast between the deep-cooperation process-innovation regions of Sunnmøre (shipbuilding) and Jæren (mechanical engineering) on one hand, and the arm's length product-innovation region of Horten (electronics) on the other. These two types of regions, as the authors pointed out, emphasize very different forms of human capital. In Sunnmøre and Jæren, industries put value on workers' informal knowledge and social qualifications, "constituted by a combination of place-specific experience, tacit knowledge and competence, artisan skills, and R&D-based knowledge" (p. 81). Such tacit skill-based knowledge, for example, was recognized as crucial by the Jæren region's ABB Flexible Automation, one of Europe's most competitive suppliers of car painting robot technologies, which alone captures around 70 percent of the European market. In Norway's arm's length Horten region, by contrast, firms place greater weight on workers' scientific, formal, knowledge. In fact, innovation cooperation in Horten is shallow precisely in the sense that their innovators only work together to the extent they possess similar formal and scientific knowledge. On a more systematic level, Busemeyer and Jensen's (2012) analysis of Eurobarometer survey data across thirteen European countries also reflects the micro mechanism behind why overeducation in academic degrees is less pandemic in better coordinated economies. When workers and students come from a "stratified" educational system that clearly differentiates "vocational training" from "academic education" (as is the case for many European countries), academic education has a self-reinforcing momentum, as more years of education further intensifies the individual's preference for academic education. Meanwhile, these "educationally stratified" European countries also happen to be "strategically coordinated" economies in the varieties of capitalism framework. Coordination, as the authors found, has precisely the opposite impact compared with stratification: while the latter intensifies preferences for academic education, the former dampens them. Furthermore, the negative impact of coordination is, strikingly, strong enough to *more than offset* the positive impact of stratification, so that, in net consequence, worker and students in more coordinated economies express *weaker* preferences for academic education. This individual level preference for academic degrees, of course, is the key micro mechanism behind the overeducation problem examined in this chapter.

Hypothesis H1, in other words, was a portrait of educational inequality from the perspective of the individual worker: it captures workers' incentives to obtain well-above-average academic education in response to the nature of coordination in their economies. Overeducation *of the worker* is the micro foundation for the macro outcome of educational "inequality from the top" *in the economy*. Unlike earnings, which have no upper boundaries, academic attainment is naturally bounded by the highest academic degree built into the system. For this reason, unlike p90/p50 in earnings inequality, a similar "measure of gap" between the top and median degrees captures solely an artifact of the educational system, and says nothing about how often this gap is turned into reality (i.e. the percentage of workforce that actually end up more educated than average in their occupation). In order to more precisely capture this macro outcome of inequality, I now test a direct macro implication of hypothesis H1:

H1 (macro implication): the percentage of workforce overeducated academically should be smaller in more coordinated capitalism.

In order to test this macro implication from H1, I aggregate the data from the level of individual survey respondents to the level of occupation (ISCO two-digit), and calculate the percentage of workforce with academic attainment well above average (i.e. two standard deviations or more) for each occupation, as the dependent variable. The more workers pull well ahead of the rest in their occupation academically, the more prevalent is such top-driven educational inequality in the economy. Because (unlike survey respondents) the same occupation is observed repeatedly across multiple years, I can now pool across the seven survey years for time-series cross-sectional analyses using fixed effects vector decomposition estimation, with Prais–Winsten adjustment for serial correlations. I report the findings from these analyses in Table 4.4.

In Model 1 of Table 4.4, all countries that appeared in the ISSP surveys are included in the analysis. The coefficient for coordination (measured using the Hicks–Kenworthy index) is statistically significant, and negatively signed. In other words, the more coordinated the economy, the smaller the percentage of workforce who have academic education well above the occupational average. A one-unit increase on the Hicks–Kenworthy index, which corresponds to the difference between minimum and maximum coordination, is sufficient to reduce the percentage of overeducated workforce by 7.41. Across the countries in the analysis, the mean percentage of overeducated workforce is 12.98. In other words, this effect reduces the instances of overeducation by more than 57 percent of its average prevalence, again a substantively notable impact. This implies that, compared with minimum coordination, there are 29.07 percent fewer instances of well-above-average education in an economy with average coordination. With all dummy control variables held at the reference category and the continuous control variable (skill level of occupation) held at its mean, the percentage of workforce overeducated for an occupation in a country with average coordination is about 10.41. A standard deviation more coordination will reduce the occupation's percentage of overeducated workers to 8.11, and two standard deviations will drive this percentage down to 5.81. Moving in the other direction, by contrast, weak coordination will make the top-driven inequality in academic attainment more pervasive in the workforce: one standard deviation below average coordination will raise the percentage of overeducated workers to 12.71, and two standard deviations will push the percentage to 15.01.

A comparison between different specific countries may provide a more concrete description for the effect of coordination in reducing the prevalence of "inequality from the top" in education. For example, in strongly coordinated Japan (0.77 on the Hicks–Kenworthy coordination index), the percentage

Table 4.4. Percentage of Workforce Overeducated for Given Occupations (ISCO88 Two-Digit), 2000–6

	(1) All countries	(2) Australia/NZ dropped	(3) Italy/France dropped	(4) Nordic dropped	(5) Hall-Gingerich measure of coordination
Coordination	−7.41(2.65)***	−6.97(1.58)***	−7.13(2.55)***	−9.90(2.54)***	−5.03(0.98)***
High specificity of vocational education	−2.95(0.93)***	−2.74(0.88)***	−3.69(0.63)***	−1.89(0.64)***	−3.75(0.89)***
Moderate specificity of vocational education	−4.84(0.57)***	−3.15(0.83)***	−2.97(71)***	−1.65(0.31)***	−2.56(0.59)***
High stratification of education	−4.60(2.0)**	−3.85(1.57)**	−4.08(2.14)	−4.35(1.89)**	−6.21(2.37)**
Moderate stratification of education	−5.21(0.28)***	−3.59(0.35)***	−3.59(0.46)***	−5.79(0.22)***	−4.70(0.98)***
High standardization of academic education	2.21(1.04)*	2.60(1.14)*	1.99(0.98)*	2.10(0.99)***	3.04(1.78)
Skill level of occupation	4.43(0.11)***	2.38(0.58)***	5.31(0.55)***	2.40(0.70)***	3.07(0.26)***
Constant	11.27(1.48)***	10.56(2.97)***	6.74(1.01)***	9.76(2.35)***	9.50(1.63)***
R-squared	0.06	0.06	0.05	0.07	0.05
N	2199	1901	1972	1599	2199

* $p<0.1$; ** $p<0.05$; *** $p<0.01$

of workforce whose academic attainment is two standard deviations above the mean is smaller by 5.63 than weakly coordinated US (0.01 on Hicks–Kenworthy). Similarly, compared to weakly coordinated UK (0.10 on coordination), strong coordination in Denmark (0.70 on Hicks–Kenworthy) will reduce by 7.37 the percentage of workers advancing far in excess of their occupation average. On the other hand, an occupation's percentage of overeducated workers in moderately coordinated France (0.40 on Hicks–Kenworthy) and Italy (0.44 on coordination) is larger than densely coordinated Norway (0.96 on coordination) by respectively 4.14 and 3.85. In short, the macro (occupation-level) outcome of "inequality from the top" in education is indeed more prevalent in less coordinated economies.

As noted in earlier chapters, the literature on varieties of capitalism has drawn increasing attention to the various "subtypes" of capitalism that exist within either liberal market or strategically coordinated economies. Some of these subtypes, such as Southern Europe or the Antipodes, are hybrid in nature, mixing symptoms of both strong and weak coordination (Hall and Gingerich 2009). One cause for concern is that the inclusion of such hybrid regimes may have biased the outcome: coordination in such countries may be both too weak (to sustain deep training) and too strong (to encourage job market signaling using academic attainment), in effect severing the link between coordination and the outcome in education. If so, when these cases are removed, the coefficient for coordination should strengthen sharply. Therefore, in Models 2 and 3, I drop respectively Australia/New Zealand and Italy/France from the analysis. The findings are not notably altered by this procedure. In other words, these hybrid cases did not disproportionately affect the outcome.

Along the same vein, I further consider another subtype on the "upper bound" of coordination: Nordic countries. In these countries, labor's organizational and political strength further enhances coordination capacity, setting them apart from continental European countries (Pontusson 2005; Amable 2003; Steinmo 2010; Ornston 2012; Kristensen and Lilja 2011). For this reason, the inclusion of these "upper bound" cases may also have disproportionately contributed to the outcome. If so, when they are removed, the effect of coordination should notably weaken. I check for this possibility in Model 4, and the finding is again not significantly altered. These various robustness checks indicate that the effect of coordination is not driven by a few countries with particular scores on coordination, but instead present across the broad range of coordination, as well as across different subcategories of capitalism. As a further check for the robustness of findings, Model 5 repeats the analyses from Model 1, but using the Hall–Gingerich index instead of Hicks–Kenworthy to measure the strength of coordination. Despite this change in the coordination measure, the core pattern of finding persists: the less coordinated the economy, the more prevalent the educational "inequality from the top" in the economy.

Such race towards top academic attainment in liberal market economies also affects the course of their technological innovation, driving R&D toward more radical technologies. Radical innovation, in turn, fuels another form of "inequality from the top," reflected in a wider top/median earnings gap. Earnings inequality will be the focus of a later section in this chapter. For now, I provide a brief mezzanine section that illuminates the causal mechanism through which innovation connects "educational inequality from the top" to "earnings inequality from the top." In particular, I show that overeducation intensifies radical innovation, which, in turn, is a powerful driver of top/median gap in earnings.

Mezzanine: Does Overeducation Boost Radical Innovation?

Because liberal market Anglo-Saxon economies have a larger percentage of their workforce overeducated, these countries may face greater distortion in the allocation of their human capital resources. In particular, more workers will be matched to job postings that on average require less academic education than they have attained, and as a result more academic knowledge will end up being underutilized relative to the routine tasks of their job. However, the *allocation* of existing resources is far from tapping the full potential of economic growth. As Penrose (1995) pointed out in her seminal theory of the firm, while firms' *allocation* of existing resources is static, their *search* for new technologies is dynamic. As a result, sources of static disadvantage may well become sources of dynamic advantage. While liberal market economies may prevent firms from fully tapping their workforce's academic qualifications during the routine tasks of their jobs, advanced academic education may nevertheless put their workforce at an advantage when their tasks become nonroutine, i.e. a part of the innovation process.

As Hall and Soskice (2001) suggest, the type of human capital needed for technological innovation may differ depending on whether the innovation is "radical" or "incremental." In the latter, the gradual quality and productivity improvement on the basis of existing products and techniques requires workers to have not only substantial experience with the existing production but also the ability to engage in constant learning-by-doing during production (Berggren 1992; Kenney and Florida 1993; Best 1990; Lazonick 1990). All these, as Hall and Soskice point out, require deep skill training of the workforce. Since, as the comparative political economy literature has documented extensively, Anglo-Saxon economies tend to fall behind their strategically coordinated European/Japanese counterparts in the training of deep, industry-based, skills (Crouch et al. 1999; Martin and Swank 2004; Thelen 2004; Iversen and Stephens 2008), Anglo-Saxon countries may do less well in incremental innovation. Along this vein, the literature has provided rich empirical evidence that while European and Japanese firms excel in high-quality and sometimes niche production, Anglo-Saxon firms focus more on lower-quality and lower-price products targeted at the mass consumer market (Hall and Soskice 2001; Sabel and Zeitlin 1997; Hollingsworth and Boyer 1997; Iversen and Soskice 2010).

On radical innovation, by contrast, the outcome may be somewhat different, as the drive towards very advanced academic education may put liberal market economies at an advantage. As the literature on technological innovation has long pointed out, successful innovation in a technology relies on substantial knowledge about this technology from actors located along the entire vertical chain from suppliers, research and development, to marketing, and finally end users (Mokyr 2002; Edquist 1997; Lundvall 2010; Edquist and McKelvey 2000). Radical innovation draws heavily on deep understanding of the principles of basic sciences. For example, advanced academic education in biochemistry and computer science is a core source of human capital for innovation in respectively the biotechnology and information technology sectors (Nelson 1996; Fransman 1990; Ornston 2012; Griliches 2000; Mansfield 1995; Helpman 1998).

Different types of occupations have different levels of requisite academic attainment. As stated earlier, the ILO ranks ISCO occupations into four general levels corresponding to four typical levels of academic education, based on ISCED classification. For the highest (fourth) level, the corresponding education falls into ISCED categories 6 and 7, which encompass university and postgraduate degrees. As a result, the highest academic degrees are the norm rather than exception for occupations at this level. These occupations all share a common ISCO one-digit classification of 2, which the ILO defines as "professionals," covering ISCO job postings ranging from "mathematicians (ISCO code 2121)," "physicists (2111)," "computer programmers (2132)," to "biologists (2211)" and "astronomers (2111, the same as physicists)." Needless to say, these occupations play an important role in radical innovation. However, the successful development of radical technologies for a market, be they new products or processes, also relies on the direct input or indirect feedback from diverse occupations below "professionals," such as "electronics and telecommunications engineering technicians (3114)," "chemical engineering technicians (3116)," "industrial robot controllers (3123)," "precision-instrument makers, machinery operators and assemblers (7311)," among others. All of these occupations have lower ISCO skill levels and ISCED academic education levels. Some of them, for example, are classified by ILO as "technicians and associate professionals" and have a corresponding average education in ISCED category 5 (four-year education starting at the age of seventeen, leading to a degree not equivalent to university degree). Others have lower ILO classifications and lower corresponding academic education (ISCED categories 2 and 3). Attainment of very advanced academic degrees in these occupations may indeed result in underutilization of academic knowledge when the production task is relatively routine (no radical innovation involved), but when the task becomes very nonroutine and the workforce become involved in radical innovation, innovation is more likely to be successful the deeper their understanding about the sciences behind such new technologies.

In short, the intense "job market signaling" via academic degrees in liberal market economies should boost their radical innovation, not only because more workers with advanced academic knowledge are produced, but also because such knowledge is diffused across a broader range of occupations that are important for successful radical innovation. As a result, different patterns of human capital investment in different types of capitalism should lead to different innovation outcomes: the deep training of skills in strategically coordinated capitalism should boost incremental innovation; the intense job market signaling through academic degrees in liberal market capitalism should boost radical innovation. Qualitative evidence from the case study literature does indicate that overeducation, defined in this book's sense of "above average for given occupation," is an important condition for success in radical innovation. For example, Crouch et al. (2009) studied two successful European centers of biomedical innovation, Munich in Germany and Cambridge in the UK. In both cases, the success of biotech firms had little to do with the national vocational training institutions, because the core staff are *"educated to higher levels than normally* associated with this form of training" (p. 667, emphasis added). In fact, it is attainment at the doctoral level that made the real difference to innovation success, and on this front, Munich actually outperformed Cambridge, with twice as many scientific publications produced in collaboration with public research institutes. Apart from Munich's ability to generate rich insider information and attract international venture capital (a topic discussed in Chapter 2), the top performance of its academic institutions is another reason why it beat the common expectation for coordinated capitalism and emerged as a highly competitive player in radical biotech innovation. However, as the authors pointed out, the Munich experience is only a "liberal island" (p. 667) in the otherwise strongly coordinated Germany economy, and as my earlier quantitative analysis had shown, overeducation is generally mild in such coordinated economies. For this reason, it is important to verify whether the link between overeducation and radical innovation can be generalized beyond isolated "special cases" to qualify as a relatively robust statistical pattern.

Statistically, do countries with more prevalent overeducation indeed have more radical innovation? In this brief section, I provide some evidence in support of this expectation. In order to identify sectors of radical technological innovation, I continue the approach in earlier chapters by following the OECD's classification of technologies (Hatzichronoglou 1997). The OECD classifies the extent of innovation into the following categories: high, medium-high, medium-low, and low, based on NACE (Rev.1) and ISIC (Rev.2). The radical innovation class covers two-digit NACE 30, 32, 33, 64, 72, or 73, and three-digit NACE 24.4 and 35.3, which incorporate sectors ranging from computers, communications, aerospace to pharmaceuticals. To measure a country's radical innovation as the dependent variable, I use the country's total number of patents (in thousands) granted by USPTO across these sectors, available from Eurostat's Science and Technology (Patent Statistics) database. Correspondingly, I take each country's mean percentage of overeducated workforce across occupations to capture the mean prevalence of overeducation in its economy, as the key independent variable. In addition, I control for R&D expenditure (available from OECD's Science and Technology Database), fixed capital investment (from OECD Structural Analysis database subsection STAN Industry), and defense spending (all as percentage of GDP), as well as the turnover ratio of the stock market (from Beck et al. (2000)'s Database on Financial Development and Structure). I report the results of panel corrected standard errors estimations in Table 4.5. The analysis covers fourteen countries[1] for 2000–6, the same time period for my earlier analysis of overeducation using data from ISSP.

Starting with Model 1, the coefficient for overeducation is 0.41: each increase by one in the percentage of overeducated workforce translates into 410 more patents in radical innovation. Since earlier findings (Table 4.4, first column) already showed that each unit of increase in coordination reduces the percentage of workforce overeducated by 7.41, the product of these two coefficients (–3.04) captures the indirect effect of coordination on radical innovation via the channel of overeducation: by discouraging overeducation, each unit increase in coordination reduces radical innovation output by around 3,040 patents. The coefficients for some variables (such as R&D or defense spending) are larger by an order of magnitude than others (such as overeducation or fixed capital investment). This, however, is an artifact of the large difference between these variables in their scale. For example, across the data, the mean for overeducation (aggregated to the country level) is 13.66 percent (of total workforce), but the mean for R&D expenditure is more than ten times smaller at 1.22 percent (of GDP). As a result, "one unit" of change

---

[1] These countries are Australia, Belgium, Canada, Denmark, Finland, France, Germany, Italy, Japan, the Netherlands, Norway, Sweden, the UK, and the US.

Table 4.5. The Effect of Overeducation on Patents from Radical Innovation, 2000–6

Model	1	2	3	4	5
Percentage of workforce overeducated	0.41(0.19)**	0.41(0.18)**	0.38(0.20)*	0.43(0.20)**	0.41(0.19)**
R&D expenditure	3.24(1.41)**	1.52(1.45)	4.1(1.47)***	9.40(2.5)**	9.14(2.37)***
Fixed capital investment	0.66(0.28)**	0.62(0.27)**	0.68(0.29)**	1.05(0.31)***	0.39(0.38)
Defense spending	9.18(1.27)***	8.01(1.19)***	10.45(1.08)***	11.10(1.39)***	9.87(1.39)***
Stock market turnover ratio	0.04(0.02)*	n.a.	n.a.	0.04(0.03)	0.06(0.03)**
Stock market liquidity	n.a.	0.06(0.02)***	n.a.	n.a.	n.a.
Stock market capitalization	n.a.	n.a.	0.001(0.01)	n.a.	n.a.
Government R&D	n.a.	n.a.	n.a.	34.56(9.66)***	34.91(9.13)***
Left party share of cabinet portfolio	n.a.	n.a.	n.a.	n.a.	−0.07(0.02)***
Constant	41.5(8.01)***	36.69(7.01)***	40.57(8.16)***	86.62(13.64)***	68.71(14.48)***
R-squared	0.61	0.65	0.59	0.70	0.74
N	77	77	77	61	61

* $p<0.1$; ** $p<0.05$; *** $p<0.01$

may imply very different scales for different variables. An increase by one in the percentage of overeducated workforce is barely a *fourth* of its standard deviation (3.83), but an increase by one in R&D as a percentage of GDP is almost *twice* its standard deviation (0.57). The effect of "one standard deviation increase" from these two variables, by contrast, is much closer in size (1,570 patents and 1,848 patents respectively).

In the remaining four models, I check the robustness of the finding using alternative specifications of various control variables. While Model 1 measured the "efficiency" of risk capital through the stock market turnover ratio, Models 2 and 3 turn to respectively the volume of trade (i.e. liquidity) and depth (i.e. capitalization) of stock markets. Besides procurement from the defense industry, the state also invests directly in R&D, in order to provide various public goods and services for the community, such as health, environment preservation, communication, and transportation. This may generate an important knowledge pool for compatible science-based innovation in the private sector (Nelson 2005; Freeman and Soete 1997). To incorporate this factor, in Model 4 I further control for R&D in the government sector. In Model 5 I add a measure of government partisanship to incorporate Roe (2002)'s finding that the redistributive demands of social democratic governments may weaken the entrepreneurial incentives of enterprises. To measure government partisanship, I use left party share of cabinet portfolios, available from Duane Swank's database Electoral, Legislative, and Government Strength of Political Parties by Ideological Group in Capitalist Democracies, 1950–2006. Very consistently across these different specifications, overeducation serves to enhance radical innovation. Depending on the specification, each one percent of the workforce overeducated translates into somewhere between 380 and 430 more patents in high-tech innovation. This implies that, by reducing the use of academic degrees for job market signaling, coordination may reduce high-tech innovation output by somewhere between 2,816 and 3,186 patents. In other words, strategically coordinated capitalism indeed makes innovation less radical, through the channel of suppressing overeducation. Having now established how overeducation feeds into patterns of technological innovation, the next section of the chapter draw outs the full implications for innovation and earnings inequality.

Whose Innovation Is More Inegalitarian?

Does technological innovation increase or reduce income inequality? The effect of economic development on inequality motivates one of the most influential theories in economic growth: the Kuznets hypothesis (1955, 1963). According to Simon Kuznets, income inequality has an inverted-U relationship with economic development. Growth widens inequality in the initial stages of economic development, but after a sufficient period of economic modernization, further growth will narrow inequality. The core mechanism proposed by Kuznets for this outcome is the economy's transition between two different sectors with large income difference (respectively agriculture and manufacturing). In the early stages of growth, a still predominantly rural economy starts to make way for the entry of a nascent manufacturing sector. During this initial period when manufacturing is still a radical innovation to the extant method of economic production, the entry of manufacturing creates various conditions that worsen inequality in society. For example, skill and income differentiation is much larger in manufacturing than in the agricultural sector which it is replacing. Consequently, the inequality of earnings rises as the workforce start to relocate from the less inegalitarian rural to the more inegalitarian urban areas.

This growing inequality is not only driven by larger income gaps *within* manufacturing, but also by a widening gap *between* manufacturing and agriculture: because of its radical improvement to the traditional method of production, manufacturing not only enjoys higher *levels* of income but also larger *increases* in income, allowing it to pull further ahead of agriculture, which loses market share to the invading new sector (Mokyr 1990). In other words, not only are new sources of income *created* during the entry of manufacturing as a radically innovative production paradigm, this process also causes income to be *redistributed*, away from the old sector to the new sector, which further exacerbates inequality. As industrialization further proceeds, however, it starts to create conditions that favor the reduction of inequality. For example, as the weight of manufacturing workers increases in the economy, a more centralized labor movement with more bargaining power emerges. Similarly, the rising social needs concomitant with modernization (old age care, education, daycare, and unemployment) spur governments to expand the generosity of the welfare state (Kuznets 1955, 1963; Wilensky and Lebeaux 1958; Kerr et al. 1960; Pryor 1968). As a result, inequality declines as more income is *redistributed* to the poor through taxes and welfare state transfers. The Kuznets hypothesis of inequality, in other words, sheds light on the *redistribution* of wealth that accompanies the *creation of new wealth* (in manufacturing) during economic development. Building on this insight from Kuznets, I now develop a theory of how innovation affects earnings inequality in advanced industrialized countries.

As the previous chapter's discussion of product and process innovation has shown, new technologies allow the innovating sector to earn rents (income above cost) from the rest of society. In product innovation, new products allow innovating firms to set prices above cost, commanding a "novelty premium" based on their temporary product monopoly. In process innovation, new methods of raising productivity allow innovating firms to push marginal costs below competitive prices, through smaller overhead, faster production, greater efficiency and flexibility, among others (Edquist et al. 2001; Chaney et al. 1991; Doukas and Switzer 1992). Both product and process innovation pass new economic resources to the rest of society, but the innovating sector is more than compensated for its cost of creating these resources: it also collects from society the "novelty rents" of product innovation and "productivity rents" of process innovation. As Aghion et al. (2002) point out, these rents allow workers in the innovating sector to earn a "human capital premium" for their ability to adapt to the needs of innovation. As a result, besides the obvious social benefit of creating new wealth, innovation also has a redistributive effect, transferring some wealth from the rest of the society to workers in the innovating sector in the form of rents. Innovation, in other words, is "creative redistribution."

A radical innovation brings to the economy a drastically new product or a drastically superior method of improving the production process. Consequently, very large novelty rents (from new products) and productivity rents (from new production processes) flow to this radical entrant sector. Because of the significant extent of innovation, this invading sector will collect larger rents than existing sectors where innovation is more mature and incremental. On one hand, the larger size of rents collected by this radical entrant sets its income at a higher level than incumbent innovators; on the other hand, the extraction of these rents from the rest of society in effect transfers wealth from incumbents to the entrant. In other words, income is redistributed from moderate-income incumbents to the high-income entrant. As a result, radical innovation increases inequality, by pulling the top ahead of the rest. This inequality-increasing effect from the entry of a radical technology mirrors the first half of Kuznets' curve, where the emergence of manufacturing as a radically new production paradigm in a still predominantly agricultural economy increases inequality.

Radical innovation, in other words, widens inequality by pulling the top ahead of the median, which may be captured as the ratio of earnings by workers at the ninetieth percentile over those at the fiftieth percentile (p90/p50 ratio). Since there is less radical innovation where there is less overeducation (i.e. strategically coordinated capitalism), coordination will also soften the inegalitarian impact of innovation.

H2: the more coordinated the economy, the smaller the marginal impact of innovation in increasing the p90/p50 earnings gap.

In comparing wage equality in the US with Europe, Acemoglu (2003b) suggests that innovation plays an important role in explaining why wages are more equal in Europe. According to Acemoglu, because rigid labor market institutions in continental Europe favor the centralized bargaining power of unions, firms have a strong desire to undermine workers' bargaining advantage by engaging in less innovation, which helps lessen firms' reliance on the supply of high-quality human capital. By innovating less, firms put lower premium on workers' human capital. As a result, the income of high-quality workers rises less sharply, which constrains the extent of wage inequality among the workforce. Although Acemoglu is similar in spirit to this chapter in implying that European innovation should create less inequality, these two accounts have some important differences. First, they seek to explain different outcomes. While Acemoglu analyzes earnings inequality as a society-level outcome, this chapter makes an explicit distinction between inequality "from the top" and "from the bottom." In fact, as I will argue below, while innovation causes *less* inequality "from the top" in Europe (concurring with Acemoglu), it actually causes *more* inequality "from the bottom" in Europe (contrary to Acemoglu). Second, the two accounts also offer different explanations. Acemoglu's theory focuses strictly on *wage bargaining* regimes, highlighting in particular the *conflict* of interest between labor and firms. This chapter, by contrast, focuses on *firm-centered* relationships (interfirm, firm–investor, and firm–labor), highlighting the role of *coordination* instead. Coordination deepens skill training, which, as I showed earlier, defuses the "trigger for educational inequality," i.e. job market signaling using academic degrees. In order to highlight the distinction between this argument and Acemoglu, it is helpful to control for bargaining centralization in the later empirical analysis. This will shed light on whether the inequality effect of firm-centered coordination remains, after the inequality effect of wage institutions is parsed out.

Although technological innovation generates less inequality from the top in strategically coordinated capitalism, this does not imply that innovation in these economies is always conducive to wage equality. In fact, as I develop in the next hypothesis, these economies allow technological innovation to widen another type of inequality: *from the bottom*. Innovation in strategically coordinated capitalism, in other words, has a *long-tail effect* on inequality.

So far, my discussion about innovation and inequality has focused on the comparison between the high-income radical innovation sector and the moderate-income incremental innovation sector. By contrast, the low-income labor-intensive noninnovative sector offers a view of inequality "from the bottom." Unable to produce technological innovation on its own, the stagnant labor-intensive sector (such as retail and personal services) is a customer of not only radical technologies (such as using specially designed scheduling software to arrange staff shifts in hotels or restaurants) but also incremental technologies (such as cutting retail staff by upgrading vending machines into multifunctional automatic kiosks) (OECD 1996b). As a result, the incremental innovation sector can earn innovation rents off the stagnant sector, in effect redistributing wealth from the low- to the moderate-income sector. This pushes the median ahead of the bottom, widening the p50/p10 gap in earnings.

By contrast, the net impact on p50/p10 from the entry of a radical innovation will be more ambiguous. The radical entrant will *widen* p50/p10 *from below* by extracting innovation rents from the low-income labor-intensive sector. However, this effect is offset by a second force in the opposite direction: because the radical entrant also redistributes income away, in the form of innovation rents, from the incremental sector, it will *narrow* p50/p10 *from above*. As a result, radical innovation will be less potent than incremental innovation in widening the median/bottom earnings gap. This has direct implications for the comparison of capitalisms in p50/p10. Because innovation is more incremental in strategically coordinated capitalism, the effect of innovation in widening p50/p10 will also be stronger in such countries. In other words, between the median and bottom earners, innovation in strategically coordinated capitalism becomes a source of *inequality* in earnings.

> H3: *the more coordinated the economy, the larger the marginal impact of innovation in increasing the p50/p10 earnings gap.*

Before proceeding to empirical analysis, it is important to clarify how this chapter used Kuznets' insight to develop its own hypotheses on innovation and inequality (H2 and H3). At its core, the Kuznets thesis (1955) predicts that inequality will first rise with the entry of a radical technology and then fall back as this technology gradually matures (i.e. an inverted-U "Kuznets curve"). For this reason, the most faithful testing of Kuznets' prediction should be carried out *over time*, observing the *full lifecycle* of a technology as it goes from radical to mature. However, in this book's setting of twenty-first century affluent capitalist economies, today's sectors of leading-edge innovation (such as biotechnology and telecommunications) are still far from entering their sunset, mature, stage. In other words, the "second (mature) half" of the Kuznets curve for today's radical innovation cannot yet be observed empirically. Given this fundamental data problem, if one intends to stay with the most direct interpretation of Kuznets, the empirical analysis would have to move to a much older time period, observing the inequality implications from the initial entry and later maturity of an older sector (such as manufacturing). Although such a study of the Industrial Revolution is an undoubtedly important task, it falls outside the book's scope of twenty-first-century affluent capitalist economies.

While Kuznets' insight is an important building block for this chapter's hypotheses, merely replicating the Kuznets hypothesis is hardly the chapter's main purpose. Instead, its purpose is to derive some novel implications for how innovation affects inequality differently for different varieties of capitalism. For this reason, I refrain from a digression into the Industrial Revolution, and focus instead on the contemporary period. Since the mature, incremental, stage of contemporary radical innovation is yet to be observed, this chapter adopted an alternative strategy of *cross-sectional* comparison (contemporaneously between countries active in radical technologies and countries active in incremental technologies), instead of a *cross-time* comparison (between active and incremental stages of the same technology). This may be a less direct reflection of the Kuznets thesis, but it is a more direct reflection of Kuznets' implications for how contemporary capitalisms differ in the impact of innovation on inequality. More broadly, such cross-country comparison *across contemporary capitalisms* is also the main analytical purpose of this book.

Empirical Analysis

Both hypotheses on earnings inequality (H2 and H3) predict interaction effects. For H2, the interaction effect is negative: coordination weakens the effect of innovation on "inequality from the top." For H3, the interaction is positive: coordination exacerbates the effect of innovation on "inequality from the bottom." For the dependent variable, I use the ratio of gross earnings for workers at the ninetieth pay percentile over workers at the fiftieth percentile (p90/p50), and analogously p50/p10, from the OECD Employment and Labor Market Statistics. The two core independent variables in the interaction are innovation and coordination. Consistent with the previous section, I continue to alternate between the Hicks–Kenworthy index and Hall–Gingerich index as two different measure of coordination. For the same robustness purpose, I also measure the overall extent of innovation in the economy using two different indicators, one on the input side (total enterprise R&D expenditure as percentage of GDP, available from the OECD's Structural Analysis Database), and the other on the output side (the total number of patents in thousands granted by USPTO, available from Eurostat's Science and Technology database, subsection Patent Statistics).

I also include a series of controls in the estimation of earnings inequality. I start by controlling for bargaining centralization, long recognized as an important institutional determinant of wage compression in the economy. To do so, I use Jelle Visser's five-point bargaining centralization index, which is based on Lane Kenworthy's classification of bargaining centralization. In this classification, a country is assigned a score of "five" if bargaining is economy-wide, "four" if there is mixed industry-wide and economy-wide bargaining, "three" if bargaining is industry-wide without regular pattern setting, "two" if there is mixed industry-wide and firm-level bargaining, and "one" if bargaining is fragmented at the firm level. Data for this variable are from Visser's Database on Institutional Characteristics of Trade Unions, Wage Setting, State Intervention, and Social Pacts.

I also control for the unemployment rate, because, as Pontusson (2005) points out, the level of unemployment in the economy has a strong effect in increasing the inequality of earnings. From the same angle, Rueda (2008) finds that, out of three conditions commonly understood to be important remedies against income inequality (welfare expenditure, minimum wage, and public sector employment), public sector employment has the most powerful and consistent impact in reducing wage inequality. For this reason, I also control for the rate of public sector employment, as percentage of total employed labor force. Young people, at the beginning of their careers, typically earn less than more established older workers. For this reason, Galbraith and Kum (2005) suggest that the growth rate of the population will likely increase inequality, as faster population growth increases the share of young people in the labor force. Correspondingly, I control for the percentage rate of annual population growth. I also control for foreign direct investment as percentage of GDP. The inequality effect from an increasing presence of international capital in the national economy is subject to different interpretations. On one hand, growing mobility of international capital in search of lower costs tends to reduce the organizational and bargaining power of unions (Western 1997), and in turn, increase wage inequality. On the other hand, capital internationalization may reinforce union–firm accommodation as a form of domestic compensation (Katzenstein 1985), which insures workers for their investment in deep human capital (Iversen 2005; Swank 2002). If so, wage inequality may fall rather than rise. Data for these control variables are from the World Bank's World Development Indicators, the OECD's Labor Force Statistics and Labor Market Statistics, and OECD Historical Statistics.

On the political side of control variables, Kelly (2005)'s study of income inequality in the US finds that government partisanship affects not only post-tax post-transfer inequality (in disposable income) but also pre-tax and pre-transfer inequality (in wages). In fact, Kelly argues that the partisan effect on wages may be even larger than on disposable income, because government policies can directly affect the extent of wage inequality through several channels such as the unemployment rate and the size of the public sector. Much of income redistribution, in other words, occurs through "market conditioning" by government policies. For this reason, I include as control the percentage of government cabinet portfolios occupied by center-left parties. The data for this measure is from Duane Swank's database Electoral, Legislative, and Government Strength of Political Parties by Ideological Group in Capitalist Democracies, 1950–2006.

Setting a contrast with plurality electoral systems and candidate-centered voting, Iversen and Soskice (2006) suggest that proportional representation

(PR) electoral systems and cohesive partisan labels are conditions in favor of greater income equality because they increase the office-winning chances of center-left parties, who are more willing to redistribute income. Although Iversen and Soskice focus mostly on inequality of post-tax post-transfer disposable income, the government's redistributive goals may also be put into action through various policies that affect market income, as Kelly (2005) noted earlier. As a result, both PR and cohesive parties may reduce earnings inequality. I control for whether the electoral system is proportional or plurality, based on Beck et al. (2001)'s Database of Political Institutions. I also control for the extent of personalism in voting, based on Johnson and Wallack (2012)'s Database of Electoral Systems and the Personal Vote. With data availability for all variables considered, the analyses cover fifteen countries for 1985–2001.[2] In Table 4.6, I report the results from (Prais–Winsten) fixed effects vector decomposition analyses of "inequality from the top," i.e. p90/p50.

I start with Model 1, which measures innovation through R&D expenditure and measures coordination using the Hicks–Kenworthy index. Although the main parameter of interest is the interaction between R&D and coordination, the "main effect" variable for R&D provides a first cut into the pattern of findings. The point estimate for this variable is significant and positively signed. In other words, when coordination is at its minimum (zero on the Hicks–Kenworthy index), each increase in R&D spending by 1 percent of GDP raises the p90/p50 ratio by 0.18, which is about 11 percent of the mean level of p90/p50 across countries in the analysis. However, this is very much the upper bound on the potential inegalitarian impact of innovation. As the statistically significant and negatively signed interaction effect indicates, the more coordinated the economy, the smaller the contribution of innovation to the top/median earnings gap. A one-unit increase in coordination will reduce the marginal contribution of R&D to the p90/p50 ratio by 0.21.

To illustrate this pattern of interaction, Figure 4.2 plots the marginal effect of R&D spending on the top/median earnings gap, conditional on the strength of coordination, with dashed lines as the 95 percent confidence interval. As the graph shows clearly, the impact of R&D on the top/median earnings inequality declines with coordination. When coordination is minimum at zero, 1 percent of GDP increase in R&D spending can raise the p90/p50 ratio by 0.18. This effect can also be directly recovered as the "main effect" coefficient for R&D in Model 1, Table 4.6. This inequality-increasing impact of R&D, however, diminishes steadily as coordination gathers strength. More than half of this impact will be shaved off by the point of average strength in

---

[2] These countries are Australia, Austria, Belgium, Canada, Denmark, Finland, France, Germany, Italy, Japan, the Netherlands, Norway, Sweden, the UK, and the US.

Table 4.6. The Effect of Innovation on Earnings Inequality, 1985–2001

	p90/p50				p50/p10			
	Model 1	Model 2	Model 3	Model 4	Model 5	Model 6	Model 7	Model 8
Coordination*R&D	-.21(0.03)***	-.38(0.14)**	n.a.	n.a.	0.26(0.02)***	0.49(0.24)*	n.a.	n.a.
Coordination* Patents	n.a.	n.a.	-0.21(0.03)***	-0.20(0.09)*	n.a.	n.a.	0.31(0.01)***	0.12(0.05)*
R&D	0.18(0.02)***	0.16(0.04)***	n.a.	n.a.	-0.07(0.02)***	0.01(0.02)	n.a.	n.a.
Patents	n.a.	n.a.	0.09(0.01)***	0.40(0.11)***	n.a.	n.a.	-0.17(0.01)	0.53(0.11)***
Coordination	0.14(0.04)***	0.12(0.06)*	0.03(0.02)	0.32(0.12)**	-0.91(0.04)***	-0.22(0.09)**	-0.34(0.01)***	51(0.36)
Foreign direct investment	0.003(0.001)**	0.006(0.002)**	0.005(0.001)***	0.008(0.002)***	0.002(0.001)	0.006(0.003)**	0.002(0.001)*	0.003(0.001)***
Public sector employment	-0.01(0.00)***	-0.02(0.00)***	-0.02(0.00)***	-0.01(0.00)***	0.002(0.00)***	0.004(0.00)***	-0.006(0.00)***	0.002(0.00)***
Left party share of cabinet portfolio	0.00(0.00)	0.00(0.00)	0.00(0.00)	0.00(0.00)	0.00(0.00)	0.00(0.00)	0.00(0.00)	0.00(0.00)
Bargaining centralization	-0.003(0.004)	-0.003(0.003)	-0.01(0.00)**	-0.009(0.004)*	0.004(0.003)	0.008(0.006)	0.001(0.002)	0.004(0.002)
Unemployment	-0.009(0.002)***	-0.007(0.001)***	-0.003(0.001)**	-0.002(0.00)***	-0.006(0.001)***	-0.009(0.002)***	-0.004(0.001)***	-0.01(0.002)***
Proportional representation	-0.06(0.04)	-0.04(0.06)	0.01(0.03)	0.01(0.02)	0.08(0.03)***	0.08(0.04)*	0.09(0.01)***	0.06(0.04)
Personalism in voting	0.03(0.02)	0.06(0.04)	0.03(0.01)**	0.04(0.02)*	0.02(0.01)*	0.06(0.03)*	0.05(0.01)***	0.02(0.01)
Annual growth in population	-0.003(0.01)	0.002(0.002)	0.008(0.01)	0.001(0.001)	0.01(0.01)	0.005(0.004)	-0.000(0.001)	-0.004(0.002)
Constant	1.79(0.09)***	1.20(0.18)***	1.83(0.08)***	1.65(0.03)***	1.89(0.06)***	1.10(0.25)***	1.68(0.04)***	1.36(0.29)***
N	179	179	168	168	179	179	168	168
R-squared	0.93	0.86	0.90	0.89	0.97	0.90	0.97	0.89

* p<0.1, ** p<0.05, *** p<0.01

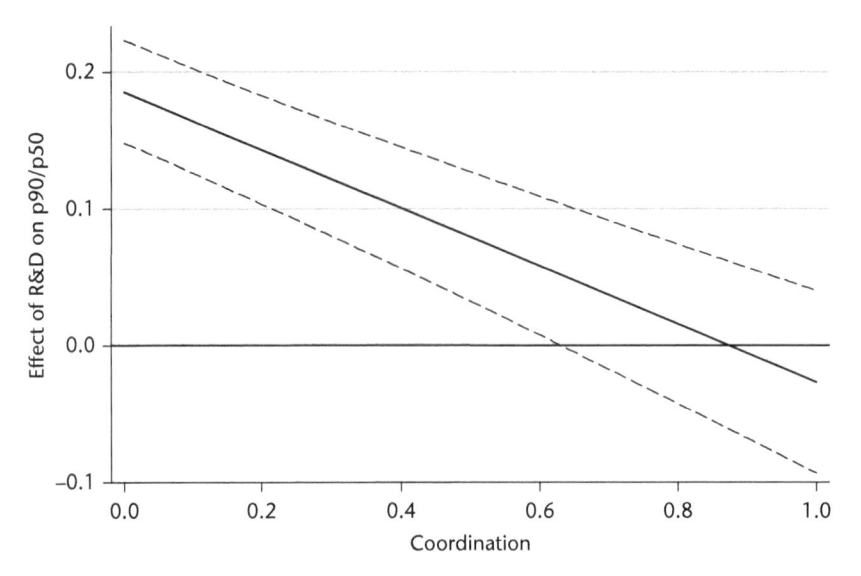

Figure 4.2. Innovation and "Inequality from the Top"

coordination (0.51 on the Hicks-Kenworthy index): at this level of coordination (above Anglo-Saxon and Southern European countries but below Northern European countries), the same increment of increase in R&D lifts p90/p50 by only 0.07. When coordination grows to around 0.62 on the Hicks-Kenworthy index (close to Belgium and the Netherlands), the effect of R&D in widening the top/median earnings gap drops out of statistical significance. In other words, when coordination is reasonably strong, technological innovation no longer has any real effect in pulling the top ahead of the rest in earnings.

To better appreciate the substantive implications of such interaction, I follow the procedure by Kam and Franzese (2007) and calculate the marginal effect (and associated standard errors) of R&D on p90/p50 for different countries. For example, while each 1 percent growth in R&D in weakly coordinated UK (0.10 on the coordination index) will drive up the top/median earnings ratio by 0.17 (standard error 0.02), the same increment of R&D increase in the Netherlands (0.58 on coordination) has only about a third of the effect, lifting the p90/p50 ratio by 0.06. Furthermore, with standard error at 0.03, this effect only clears statistical significance at the 0.1 level. Similarly, the inegalitarian impact of innovation in strategically coordinated Denmark (0.72 on coordination) is notably smaller than liberal market US (0.02 on coordination): while each 1 percent R&D increase in the US raises the top/median earnings ratio by 0.18, the impact from the same R&D increase in Denmark is six times smaller, raising p90/p50 by only 0.03. With the standard error for this marginal effect at 0.03, it is clear that in reality R&D in Denmark has no statistically significant impact in increasing "inequality from the top."

The interaction effect can also reveal differences in p90/p50 between countries that are less drastically different from each other in their strength of coordination. For example, between two of the most important continental European economies (Germany and France, respectively 0.80 and 0.40 on the Hicks–Kenworthy coordination index), while 1 percent R&D increase in the former has no statistically significant effect on the top/median earnings gap, the same R&D increase in the latter pushes p90/p50 up by 0.10. Furthermore, the inegalitarian effect in France comfortably clears significance at the 0.01 level, with standard error being 0.02. In short, evidence from Model 1 supports H2's prediction that innovation causes less "inequality from the top" in more coordinated capitalism. Model 2 repeats the analysis of Model 1, this time using the Hall–Gingerich index to measure coordination. The pattern of finding remains unchanged: the interaction between R&D and coordination is negatively signed, and significant at the 0.01 level. At minimum coordination (zero on the Hall–Gingerich index), each increase in R&D spending by 1 percent of GDP adds 0.24 to the p90/p50 ratio. However, as coordination strengthens, this inegalitarian impact of R&D also weakens. Moving from minimum to maximum coordination (i.e. one unit of increase on the Hall–Gingerich index) will reduce the marginal contribution of R&D to the p90/p50 ratio by 0.38. In other words, compared to minimum coordination, each standard deviation increase in coordination (0.29 units on the Hall–Gingerich index) will slash the inegalitarian impact of R&D by around 46 percent.

Instead of R&D spending, Model 3 measures innovation through the total number of patents (in thousands) granted by USPTO. The key variable of interest is therefore the interaction between patents and coordination (measured on the Hicks–Kenworthy index). To set the stage for the interaction, we can again start with the "main effect" coefficient for patents, which captures the marginal effect of patents on "inequality from the top" when coordination is minimum at zero. Under this circumstance, each thousand new patents add around 0.09 to the p90/p50 ratio. This, however, is again the upper bound on the inegalitarian impact of new patents, because patents interact negatively with coordination in their effect on inequality, and the interaction comfortably clears significance at the 0.01 level. To see how this interaction unfolds, Figure 4.3 plots the marginal effect of patents on "inequality from the top," conditional on the strength of coordination.

As the graph shows clearly, the impact of patents on "inequality from the top" shifts with coordination. When coordination is minimum at zero, each thousand new patents add around 0.09 to the p90/p50 ratio, but this inegalitarian effect declines quickly as coordination gains strength. The effect drops out of statistical significance after coordination reaches around 0.20 on the

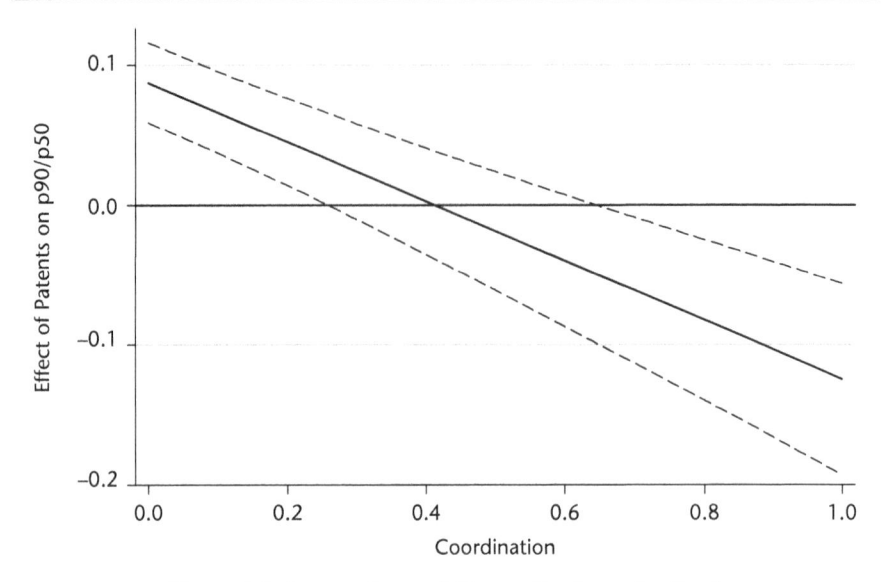

Figure 4.3. Innovation and "Inequality from the Top"

Hicks–Kenworthy index, which neatly separates Anglo-Saxon countries from other advanced industrialized economies. In other words, while new patents imply larger "inequality from the top" in Anglo-Saxon countries, this is not the case for the rest of the advanced industrialized world. In fact, as the graph shows, when coordination becomes reasonably strong (above 0.63 on the coordination index, covering most Northern European countries except the Netherlands), the effect of new patents on p90/p50 *flips from positive to negative*. In other words, for these countries, technological progress *narrows* the top/median earnings gap. Model 4 repeats the analysis of Model 3, substituting the Hall–Gingerich measure of coordination for the Hicks–Kenworthy measure. The core pattern of finding remains, though now the interaction clears significant at the 0.1 level.

On the whole, the findings so far suggest that weak strategic coordination results in a *superstar effect*, i.e. "inequality from the top." When coordination is weak, innovation has a statistically significant and substantively notable effect in pushing the top ahead of the median in earnings. This pattern is found when innovation is measured as an input (R&D) as well as output (patents), for coordination measured either on the Hicks–Kenworthy or Hall–Gingerich index. The more coordinated the economy, the smaller this *superstar effect*. In very strongly coordinated capitalism, this *superstar effect* completely vanishes: in these countries, innovation has no statistically significant effect in widening "inequality from the top."

Nevertheless, the absence of the *superstar effect* in strategically coordinated capitalism does not imply that innovation in such economies is always good

Whose Innovation Creates More Inequality?

for wage equality. In fact, as H3 predicted, technological innovation in these economies drives another type of inequality, from the bottom (i.e. the *long-tail effect*). I analyze such "inequality from the bottom" (i.e. p50/p10) in the remaining four columns of Table 4.6. Here, according to H3, coordination should now exacerbate the effect of innovation in widening the median/bottom earnings gap. This prediction is borne out in Model 5, which measures innovation using R&D expenditure and measures coordination through the Hicks–Kenworthy index. In Model 5, the interaction coefficient is positively signed and clears significance at the 0.01 level. In other words, the inegalitarian impact of innovation on "inequality from the bottom" is stronger in more coordinated capitalism. Based on this estimation, Figure 4.4 plots the marginal effect of R&D on p50/p10, conditional on the strength of coordination. As the graph shows, when coordination is at a minimum level, R&D reduces rather than increases the median/bottom earnings gap. However, as the economy becomes more coordinated, the effect of R&D soon flips over, and becomes steadily more inegalitarian. At mean coordination, each 1 percent of GDP increase in R&D spending will raise the p50/p10 by around 0.07. When coordination edges towards maximum (close to Sweden and Norway), the same R&D expansion will raise p50/p10 by 0.17, which is more than 2.5 times the lifting power delivered at mean coordination.

To check the robustness of this finding, Model 6 repeats the analysis of Model 5, this time using the Hall–Gingerich measure of coordination instead of Hicks–Kenworthy. As the sixth column in Table 4.6 shows, the qualitative

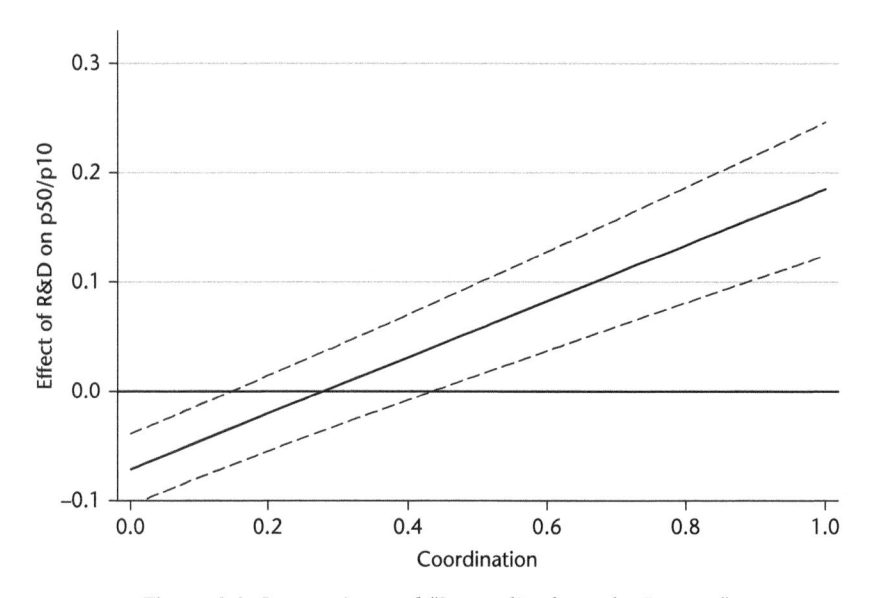

Figure 4.4. Innovation and "Inequality from the Bottom"

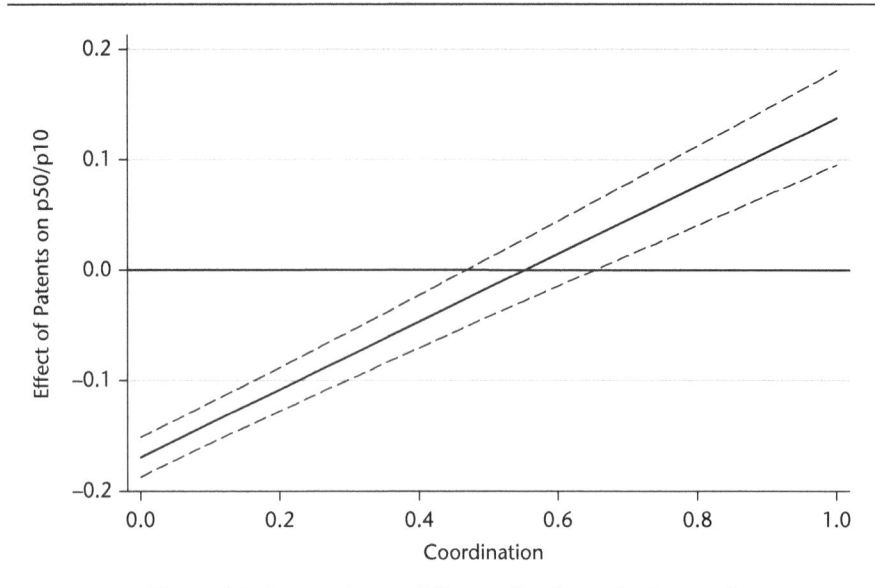

Figure 4.5. Innovation and "Inequality from the Bottom"

pattern of finding remains unchanged: R&D interacts positively with coordination in its impact on "inequality from the bottom." Instead of R&D spending, Model 7 again measures innovation using patents (and measures coordination using Hicks–Kenworthy). The core finding again goes through: there is a positive interaction between patents and coordination. This is corroborated in the plot of marginal effects in Figure 4.5: the more coordinated economy, the stronger the impact of patents in widening the median/bottom earnings gap.

The same positive interaction effect can be found in Model 8, which re-analyzes Model 7 by substituting the Hall–Gingerich measure of coordination for the Hicks–Kenworthy measure. In short, while liberal market capitalism allows technological innovation to create a *superstar effect* (i.e. driving inequality from the top through p90/p50), innovation in strategically coordinated capitalism has a *long-tail effect*, driving inequality from the bottom through p50/p10.

Besides these quantitative findings, qualitative evidence from the case study literature further illuminates the very process of "redistribution" that underlines the liberal market "inequality from the top" uncovered in the above statistical analysis. For example, William Lazonick's (2010) examination of financialization in US corporate governance since the 1980s highlighted a massive "reallocation" of financial and human capital from "established old-economy corporations to new-economy startups" (p. 694), powered by the creation of NASDAQ and various other financial instruments geared towards radical technological innovation. The nexus of high-tech startups, venture capital, and IPOs, in turn, drove a sharp increase in stock repurchases as a primary means of redistributing rents to the new economy. Among the 438 companies in the S&P 500 index that went public in 1997, a total of $2.8 trillion was expended on stock repurchases over the next thirteen years, with an average of $6.4 billion per company. Over the same time period, a total of $2.0 trillion was distributed in the new economy as cash dividends, with an average of $4.6 billion per company. As Lazonick pointed out, the redistribution of wealth to the high-tech sector through repurchasing on the stock market not only intensified volatility (a key problem for liberal market economies as the next chapter will show), but also massively widened earnings inequality from the top. Drawing on data from AFL-CIO Executive Paywatch, Lazonick highlighted that the top/average wage ratio, understood as the ratio of average CEO pay for 200 large US corporations over the average pay of the full-time US worker, had exploded, from 42:1 in 1980 to 107:1 in 1990, reaching 525:1 in 2000. Despite losses from the recent global financial crisis, the ratio still remained at 319:1 in 2008, almost eight times the already large gap in 1980.

Besides the wide "inequality from the top" in liberal market economies, the case-study literature also offers a window into "inequality from the bottom" in coordinated economies, driven by the gap between their core incremental-innovation sectors and peripheral noninnovative low-skill sectors. For example, Martin and Knudsen (2010) found that Denmark, as an example of strategically coordinated economy, actually had a *larger* share of minimum-skill workers (44 percent) in its retail service sector in 1997 than the UK (34 percent), an example of a liberal market economy. By 2003, although the share of low-skill workers declined in both countries' retail sectors, the 10 percent gap between countries remained intact. Denmark's relative concentration of minimum-skill workforce in the low-end retail sector highlights the segmentation between the country's core incremental-innovation industrial sectors and its labor-intensive noninnovative low-end sectors, a general problem identified in this chapter for strategically coordinated economies. Nevertheless, as the authors carefully pointed out, greater core/periphery *gap of innovation potential* in the Danish economy should not be equated with greater *level of earnings inequality*. Because of various wage bargaining, labor market, and welfare state institutions that mitigate earnings inequality, the actual level of wage inequality in Denmark is lower than the UK. In 2009, for example, the Gini coefficient for the UK was 0.36, while it was only 0.25 for Denmark (Martin and Knudsen 2010, p. 360). This subtlety in understanding the meaning of "inequality" is also an overall theme emphasized in this chapter: as I noted earlier, the main outcome of interest in the chapter is not the *level* of inequality, but the *slope* between inequality and innovation. Instead of asking what countries have a wide p50/p10, so to speak, the chapter asks what countries allow technological innovation to widen p50/p10. As a result, there is no real contradiction between the finding that innovation *widens* the median/bottom earnings gap in coordinated economies and the fact that the *size* of this gap is generally smaller in these countries. Instead, they bring together different dimensions (level and slope) to paint a more complete picture of earnings inequality: while Anglo-Saxon countries suffer higher "inequality from the bottom" than Nordic Europe, new technologies may *blunt* this difference, by widening the median/bottom gap in Nordic countries.

The quantitative analyses of inequality so far have focused on wage ratios. As measures of inequality, wage ratios have the flexibility to identify inequalities specific to particular segments of the earning scale (i.e. "from the top" and "from the bottom"). By contrast, two other well-known inequality measures, the Gini coefficient and Theil's T, paint a broad picture of overall income inequality, across the economy as a whole. Because these two measures describe inequality from somewhat different angles than wage ratios, it will be interesting to draw out the implications for overall inequality from the findings on wage ratios so far. Put together, hypotheses H2 and H3 suggest that innovation in strategically coordinated capitalism will narrow the top/median earnings gap but widen the median/bottom earnings gap. In other words, innovation in strategically coordinated capitalism exerts two mutually offsetting forces on overall inequality in the economy: on one hand, it reduces inequality *from above*; on the other hand, it widens inequality *from below*. Between these two opposite forces, which one should have a greater impact on overall inequality? This should depend on whether inequality from above or below is larger in the economy. As Lupu and Pontusson (2011)'s influential study points out, inequality from above (p90/p50) is larger than inequality from below (p50/p10) in almost all advanced industrialized countries. The authors refer to this ratio of "inequality from above" over "inequality from below" as the *skew* of inequality, and they found that the *skew* is larger than one, for sixteen out of the eighteen OECD countries commonly studied in the comparative political economy literature. Because "inequality from the top" is far more extensive than "inequality from the bottom" for most countries, the effect of coordination in mitigating "inequality from the top" caused by innovation should *more than compensate* its contrary effect in exacerbating "inequality from the bottom" caused by innovation. In other words, the net impact of coordination on overall inequality in the economy should be negative: the more coordinated the economy, the smaller the effect of innovation in increasing overall inequality in the economy.

> H4: *the more coordinated the economy, the smaller the marginal impact of innovation in increasing overall inequality of earnings in the economy.*

I test this hypothesis against both measures of overall inequality noted earlier: the Gini coefficient and Theil's T. I start the analysis with the Gini coefficient, which is well known for being very intuitive in its interpretation: bounded between zero and one, it measures the proportion of income that still needs to be redistributed in order to achieve perfect equality. Both the Luxemburg Income Study (LIS) and the OECD provide data on the Gini coefficient for pre-tax pre-transfer earnings for a large number of OECD countries. Because the LIS data relies on country-specific household surveys, its time coverage is sporadic, restricted only to particular years when a survey is conducted, which differ from country to country. These highly sporadic time points are poorly aligned with the limited time points for the various control variables I use for estimating inequality. As a result, many observations are lost from missing values when I use the Gini data from LIS. By contrast, the OECD data on the Gini coefficient suffers less from this problem. The OECD's Data Set on Income Distribution and Poverty provides the Gini coefficient for the working age (15–65 years of age) population. In this data set, each country's Gini value is calculated as a "period average" for separate time periods: "mid-1980s," "around 1990," "mid-1990s," "around 2000," "mid-2000s," and "late-2000s." Correspondingly, for each country I also calculate the period averages for all other variables in the analysis, based on the following time intervals: 1983–7 for "mid-1980s," 1988–92 for "around 1990s," 1993–7 for "mid-1990s," 1998–2003 for "around 2000," and 2004–6 (no data for 2007) for "mid-2000s." As a result, all variables in the analysis are measured by five common time periods instead of annually, which prevents the creation of missing values. By taking a variable's average across a given period, I will be able to incorporate the information from all its available observations during this period, even if some of their corresponding observations on other variables are missing. In other words, a single missing value in one variable will not result in the dropping of corresponding observations in all other variables. Besides the dependent variable (Gini) and core independent variables (innovation and coordination), I also retain all the control variables used in the chapter's earlier analyses of earnings inequality. I report the results of the Gini analyses in the first two columns of Table 4.7. These analyses cover the same fifteen OECD countries as the earlier wage ratio analyses. All variables are within-period averages, calculated for five different time periods as specified earlier.

Model 1 measures innovation through R&D spending, while Model 2 measures innovation through patents. The hypothesis H4 predicts an interaction effect: the more coordinated the economy, the smaller the impact of innovation in widening overall inequality in the economy. This prediction is borne out in both models: the point estimation for the interaction between innovation and coordination is statistically significant and negatively signed in

Table 4.7. The Effect of Innovation on Earnings Inequality, Additional Analysis

	Gini Coefficient		Theil's T	
	Model 1	Model 2	Model 3	Model 4
Coordination*R&D	−0.07(0.01)***	n.a.	−0.003(0.001)***	n.a.
Coordination*Patents	n.a.	−0.02(0.007)**	n.a.	−0.005(0.002)*
R&D	0.04(0.003)***	n.a.	0.004(0.000)***	n.a.
Patents	n.a.	0.08(0.001)***	n.a.	0.012(0.003)***
Coordination	0.01(0.02)	−0.01(0.004)**	−0.003(0.000)***	−0.002(0.000)***
Foreign direct investment	0.006(0.006)	0.002(0.003)	−0.0002(0.0000)***	−0.0002(0.0000)***
Public sector employment	0.002(0.001)*	0.001(0.000)**	0.0001(0.0001)	0.0001(0.0001)
Left party share of cabinet portfolio	−0.01(0.003)***	−0.01(0.002)***	−0.0002(0.0001)	−0.0001(0.0001)
Bargaining centralization	−0.04(0.001)***	−0.08(0.003)**	−0.008(0.002)***	−0.007(0.003)*
Unemployment	0.005(0.000)***	0.002(0.000)***	0.0004(0.0001)***	0.0001(0.0000)***
Proportional representation	0.11(0.15)	0.07(0.06)	0.01(0.02)	0.006(0.004)
Personalism in voting	0.05(0.06)	0.02(0.01)	0.004(0.002)*	0.002(0.001)*
Annual growth in population	0.000(0.001)	0.001(0.001)	0.0000(0.0001)	0.0000(0.0001)
Constant	0.28(0.17)	0.20(0.09)*	0.01(0.00)***	0.02(0.00)***
N	70	68	225	228
R-squared	0.81	0.75	0.31	0.43

* $p<0.1$; ** $p<0.05$; *** $p<0.01$

Whose Innovation Creates More Inequality?

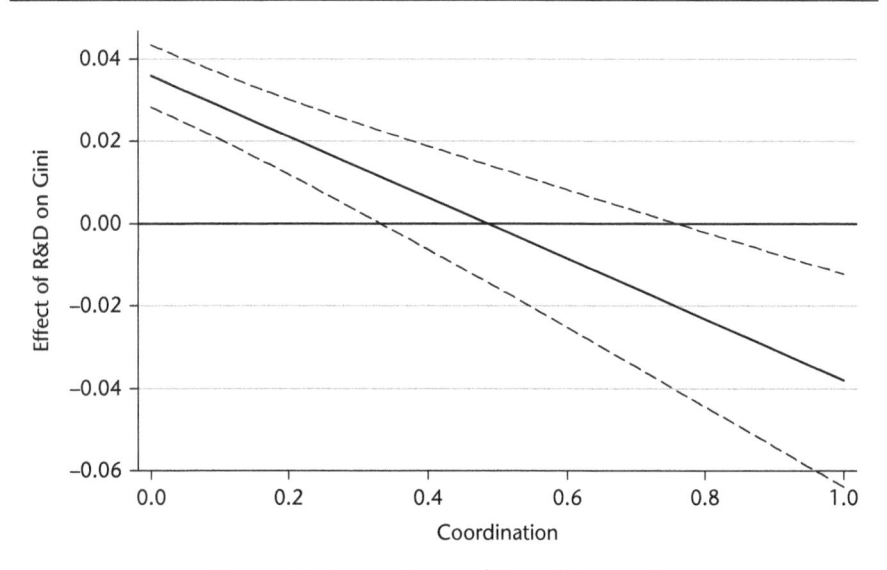

Figure 4.6. Innovation and Overall Inequality

both cases. To illustrate the interaction, I use Model 1 as an example and plot the marginal effect of R&D on the Gini coefficient in Figure 4.6, conditional on the strength of coordination (measured on the Hicks–Kenworthy index).

As the graph shows, depending on the strength of coordination, the effect of R&D on the earnings Gini coefficient can be organized into three different "regions." In the region of "weak coordination" (roughly below 0.3 on the Hicks–Kenworthy index, which neatly overlaps with the full set of Anglo-Saxon countries), innovation is a source of inequality: for these countries, R&D has a statistically significant effect in raising the Gini coefficient of earnings. For example, at the minimum level of coordination (close to the US), 1 percent of GDP growth in R&D spending will increase the Gini coefficient by around 0.04, which is about a tenth of the Gini's actual value for the US (0.42) averaged across the time period of the analysis. In the region of "moderate coordination" (roughly between .3 and .8 on the Hicks–Kenworthy index, which encompasses most continental European countries), innovation has no statistically significant impact on overall inequality: for these countries, the confidence interval straddles across the zero line. More strikingly perhaps, in the third region of "strong coordination" (above 0.8 on the Hicks–Kenworthy index, which covers most Nordic countries plus Austria), the effect of innovation on inequality flips its sign: for these countries, R&D has a statistically significant impact in *reducing* the Gini coefficient of earnings. For example, at the maximum level of coordination (close to Sweden), each 1 percent of GDP growth in R&D spending will shave about 0.04 off the Gini coefficient, in effect reducing the already low level of Swedish earnings inequality (Gini=0.34 averaged over time) by another 12 percent. In other words, in between maximum and minimum coordination lies a difference in Gini of almost 0.08, which is almost a fifth of the average Gini coefficient (0.39) across the fifteen countries in the analysis. The comparison among these three regions in the graph shows that strategically coordinated capitalism may indeed transform technological innovation from a source of overall inequality into a source of equality instead.

In the next two columns of Table 4.7, I turn from the Gini coefficient to Theil's T as an alternative measure for the overall inequality of earnings in the economy, based on data from the University of Texas Inequality Project (UTIP). Theil's T measures earnings inequality across sectors in the entire economy as a weighted sum of the extent to which any sector's average earning ranks above the entire economy's average earning. The larger the workforce in a given "above-average-earning" sector, the more heavily this sector is weighted in calculating the economy's overall inequality of earnings. Of course, this principle to "count inequality more when it affects more people" was also reflected in a previous section of this chapter, where I constructed a macro (i.e. occupation-level) measure of educational inequality based on each occupation's percentage of workforce with well-above-average academic attainment. Using industrial statistics from the United Nations' Industrial Development Organization (UNIDO), UTIP calculates Theil's T for a large number of developed and developing countries. As is well known, the raw value of Theil's T lacks an intuitive interpretation (unlike the Gini coefficient), and it can only be meaningfully interpreted in the context of "relative comparison" across different observations (i.e. "relative changes" in inequality). This, however, does not pose a real problem here, because this chapter's discussion of interaction effects naturally focuses on "changes in inequality" as the marginal effect of innovation, conditioned by the strength of coordination. In terms of coverage, the advantage of the UTIP data is that it has uninterrupted annual observations up to the year 2002 for a large number of countries; the disadvantage is that it only covers sectors related to manufacturing.

Models 3 and 4 in Table 4.7 report the findings on Theil's T, measuring innovation through R&D spending and patents respectively. These analyses cover fifteen countries[3] for a period between 1980 and 2002. Findings from both models are consistent with those from the Gini coefficient: innovation interacts negatively with coordination in its impact on overall inequality in the economy. In other words, the more coordinated the economy, the less impact of innovation in widening the overall inequality of earnings.

---

[3] These countries are: Australia, Denmark, Finland, France, Germany, Ireland, Italy, Japan, the Netherlands, New Zealand, Norway, Sweden, Switzerland, the UK, and the US.

Overall, this chapter has revealed a relatively nuanced pattern of inequality and contemporary capitalism: innovation in different types of capitalism may be associated with different types of earnings inequality. Liberal market capitalism has a *superstar effect*, increasing "inequality from the top." This is reflected not only in more workers becoming overeducated at the very top of the educational ladder, but also a wider top/median earnings gap as a result of technological innovation. By contrast, innovation in strategically coordinated capitalism has a *long-tail effect*, increasing "inequality from the bottom" through a wider median/bottom earnings gap. Furthermore, because "inequality from the top" in earnings is more extensive than "inequality from the bottom" for most advanced industrialized countries (Lupu and Pontusson 2011), the effect of coordination in mitigating "inequality from the top" *more than compensates* its contrary effect in exacerbating "inequality from the bottom." In other words, in its net effect, coordination constrains the impact of innovation on overall inequality in the economy, as my findings from the Gini coefficient and Theil's T indicate.

Conclusion

Today, the inequality of earnings is on the rise in a larger number of OECD countries, with damaging consequences for the low-paid (Pontusson 2005; Galbraith 1998; Galbraith and Berner 2001; Alderson and Nielsen 2002). Besides earnings, the bottom segment of the labor force is also hurt by the inequality of human capital attainment. In continental Europe, for example, this is reflected in labor market outsiders' poor skill training compared with insiders (Rueda 2005; 2007). In the US, this is reflected in the growing number of students falling behind academically in high school due to poverty and income inequality (Goldin and Katz 2008; Berg 2010). Instead of this common "view from the bottom," this chapter has adopted an opposite "view from the top" on inequality. Its central message is simple: liberal market capitalism suffers from two types of inequality "from the top": an educational gap (top academic attainment well above the average for given occupation), which in turn allows innovation to drive an earnings gap (top well above the median in earnings); innovation in strategically coordinated capitalism, by contrast, widens inequality "from the bottom" (bottom well below the median in earnings).

Why do some countries drive workers to attain academic education well above the average for their occupation, but other countries encourage workers to anchor to the average? This chapter has shown that overeducation at the very top of the educational ladder is especially prevalent in liberal market capitalism. Because weak coordination discourages training, workers lack skills informative of their potential productivity. As a result, they have to use academic credentials to signal productivity (Spence 1973), even if they invest in far more academic education than will be used in their actual job (Thurow 1976; Borghans and de Grip 2000b). As I have shown based on multiple waves of ISSP data, the likelihood of overeducation is higher in less coordinated capitalism. In such economies, a larger share of the workforce ends up with the very top academic attainment, well above the average used in their occupation.

Earlier in Chapter 2, I demonstrated a form of pathological competition in financial markets: rat races. Overeducation is yet another example of the rat race, this time on the job market. It may be easier to further understand this form of competitive pathology by comparing it with healthy competition. Healthy competition often fulfils two valuable objectives: (1) determining who wins (i.e. a "prize allocation effect"); and (2) pushing the frontier in possible output (i.e. a "growth effect"). For example, a sports tournament not only identifies the champion ("allocation effect"), but also motivates better performance ("growth effect"), as nicely summarized in the Olympic motto of *Citius, Altius, Fortius* (Faster, Higher, Stronger). When workers on the production line compete for a bonus pay, the final winner ("allocation effect") is also likely to have achieved the greatest increase in output ("growth effect") (Aoki 1989). Similarly, in healthy competition for good jobs, whoever emerges as the job winner ("allocation") is likely to bring more productivity to the firm ("growth"). The joint importance of these two competitive goals ("allocation" and "growth") may be better appreciated by using the well-known "growth vs. distribution" dilemma as a metaphor. As the distributive politics literature points out, redistributive concerns have to be balanced with the concern for continuing growth. Distribution without growth is self-defeating, and unsustainable in the long run (Battaglini et al. 2012; Bueno de Mesquita et al. 2003; Berry 2008; Stasavage 2005; Hicken et al. 2005; Hallerberg and Marier 2004).

The Anglo-Saxon job market rat race identified in this chapter is precisely an example of competition where there is distribution of the prize, but no growth in output. When employers make hiring decisions based on academic "credential rankings," job applicants may be *sorted* on the job market (more educated workers get better jobs), but jobs and applicants are not *matched* (workers are more educated than their jobs require). As a result, although job market competition fulfills the first objective of "prize allocation" (assigning better jobs to applicants with higher academic degrees), it crashes on the second objective of "growth": higher academic degrees do not translate into commensurate productivity improvement on the actual job. By contrast, in European and Japanese economies, dense coordination encourages training, and workers carry more substantial skills indicative of their latent productivity. This gives employers direct information to determine whether a job applicant has the productivity commensurate with the job, which in turn reduces the applicant's need to use ever-higher academic degrees as productivity signals. As a result, not only are job applicants *sorted* (better skilled workers end up with better jobs), jobs and applicants are also *matched* (workers' human capital qualifications are commensurate with the requirement of their jobs). Therefore, strategically coordinated economies may have healthier job market competition than their liberal market counterparts: competition not only allocates the prizes (better jobs to better workers) but also translates worker qualifications (skills) into real productivity growth. The finding that European/Japanese capitalism alleviates rat races on the job market adds much new relevance to Peter Katzenstein (1985)'s classic insight that coordinated capitalism avoids bad competition.

Up to this point, the book has used *markets* (financial, product, as well as labor) as the canvass to study the political economy of information distribution. The next chapter adopts a more "micro" setting, and studies how the *power to exploit information* is distributed *across the relationship* between firms during innovation. In particular, when one firm asks another to carry out R&D on its behalf, how much leeway does it grant to its partner in exploiting its own private information? As I will show, this question has direct implications for how firms cope with one of the most important economic risks: volatility.

5

Who Faces a Dilemma between Volatility and Output in Innovation?

An important motivation behind the previous chapter's focus on inequality is the potentially wide socioeconomic impact of inequality. For example, the inequality of earnings and employment opportunities is a key source of "new social risks" that accompany the transition to the postindustrial service economy. Symptoms of such new social risks include persistent unemployment for labor market outsiders, single parents working in poverty, and growing numbers of children or young couples living in poverty. These issues are gaining importance in the comparative political economy literature, especially in scholarship on welfare state and labor market policies (Armingeon and Bonoli 2006; Rehm 2011; Mares 2003; Rueda 2007; Lynch 2006; Rehm et al. 2012). However, as Jacob S. Hacker (2004) pointed out in his study of social risks in the US, an even faster-growing cause of social insecurity than the *inequality* of income is the rising *volatility* of income. For example, Moffitt and Gottschalk (2002) (cited in Hacker 2004) found that the volatility of male wages increased significantly during the 1970s and 1980s, driven more by unstable pay than by unstable job tenure. Similarly, based on studies of panel data with Nigar Nargis, Hacker (2003) suggests that the variance of family income more than doubled in the two decades since 1974, peaking in the mid-1990s at five times the level in the early 1970s.

The volatility of pay for workers is just the tip of the iceberg from a large and multifaceted literature that explores economic risks as *a spread of the variance*, in economic outcomes such as economic growth (Doucouliagos and Ulubaşoğlu 2008; Béjar and Mukherjee 2011), stock prices (Leblang and Mukherjee 2004, 2005), government revenue (Nielsen et al. 2011; Afonso and Furceri 2010), workers' earnings (Scheve and Slaughter 2004; Hacker 2004), inflation (Roberts and Wibbels 1999), and levels of consumption (Kim 2007), among many others. This literature has highlighted that the economic risks of high volatility not only have far-reaching consequences for economic performance, but also deeply shape the strategic choices of many stakeholders in society, including voters, firms, investors, and governments.

Many scholars, for example, point out that economic volatility slows down the speed of economic growth. Volatility in various public budget components (such as trade and indirect taxes, government investment expenditure, government consumption, and social security contributions) has a large and significant effect in reducing economic growth across OECD countries (Afonso and Furceri 2010). Similarly, unexpected fluctuations in money supply, exchange rate, and domestic credit are all negatively related to economic growth (Brunetti 1998). Economic volatility slows down economic growth because a spread in economic variance deters risk-averse investors from making potentially irreversible investment in capital and labor. Volatility in economic outcome may also increase the temptation for governments to call early elections opportunistically (Kayser 2005). This is because the incumbent is only willing to call an election if the current economic state will be better than what the incumbent can expect in the long-term future. When the incumbent faces a "spread in variance" in its current economic state, it becomes more likely for a random draw from this current state to exceed its long-run expected mean, and in turn more likely for an election now to be more advantageous than an election later. Because opportunistically calling an election and politically intervening in the economy (which Kayser refers respectively to as "surfing" and "manipulating") are substitute means of political survival, when economic volatility intensifies "surfing," it also decreases government "manipulation" of the economy. Economic volatility may also affect governments' incentives for manipulation in a different framework: fighting inflation. While checks and balances (i.e. multiple veto players) generally enhance government credibility in fighting inflation, such benefits disappear when the economy suffers from large volatilities (Keefer and Stasavage 2003). When there is great uncertainty regarding the economic status quo, veto players put greater weight on the government's preferred inflation outcome than the status quo. As the preference of veto players moves closer to the government, their check-and-balance influence declines.

Besides politicians, economic volatility also shapes the beliefs and strategies of various economic actors. Public sector workers, for example, are more willing to go on strikes when there is an increase in the volatility of government resources (such as oil revenue) that may be used to pay for concessions to workers (Robertson 2007). Such volatility heightens the uncertainty about the government's bargaining power, which in turn prompts workers to test the waters by engaging in more industrial action. The greater uncertainty from a spread in economic variance may also sharpen workers' sense of economic insecurity. For example, foreign direct investment increases workers' perception of economic risks because the inflow of such foreign capital adds to the volatility of both employment prospect and wages (Scheve and Slaughter 2004). Studies of economic volatility have also provided a more nuanced interpretation of Peter Katzenstein's (1985) "domestic compensation" thesis for small open economies. In these classic cases of neocorporatism, the generous provision of social security and dense coordination are commonly understood as means of compensating domestic workers for economic "openness," that is, *exposure* to the world economy (Cameron 1978, 1984). However, Kim (2007) finds that income, consumption, and investment fluctuations are actually caused by external "risks," that is, *volatility in the conditions of international trade*, such as terms of trade and exchange rates. Once *volatility* in trade conditions is parsed out, *exposure* to trade no longer has any tangible effect on these economic aggregates. In other words, instead of domestic compensation for *openness*, neocorporatist institutions may be better understood as mechanisms to smooth *risks*.

This issue of how strategically coordinated capitalism smooths the risks of economic volatility is the main theme of this chapter. Of course, a single chapter cannot examine all types of volatilities for all economic actors. Given the book's firm-centered perspective on contemporary capitalism (Penrose 1995; Hall and Soskice 2001; Pitelis 2002), this chapter focuses in particular on volatilities faced by the *firm*. This focal attention on the firm sets the book apart from studies that examine economic risks more immediately pertinent to workers (i.e. volatility in employment opportunities and wages), the government (i.e. volatility in aggregate economic outcomes, budget conditions, and policy settings), consumers (i.e. volatility in prices), or investors (i.e. volatility in stock markets) (Krause and Corder 2007; Scheve and Slaughter 2004; Hacker 2004; Doucouliagos and Ulubaşoğlu 2008; Béjar and Mukherjee 2011; Leblang and Mukherjee 2005; Roberts and Wibbels 1999; Kim 2007). In contrast to these studies, I focus on the volatility in private enterprise revenue. As profit-maximizing entities, firms have not only larger stakes but also greater control over the volatility of their own revenue stream, in comparison with the various "background" conditions such as economic growth, government spending, or the effective functioning of the stock market. Of course, firms also pay attention to the volatility of these background conditions, but unlike their own revenue, these macroeconomic conditions are unlikely to be altered by an individual firm's own strategies. Given the infeasibility of examining all types of volatilities in a single chapter, I focus on the one type that most directly and transparently motivates firms' strategies: volatility in firm revenue.

The volatility of a firm's revenue stream may be caused by many economic, social, and political factors. Again, to reflect this book's central theme of firm-centered coordination, I focus in particular on *opportunism in principal–agent relationship between firms* as the source of revenue volatility. When one firm carries out an R&D task on behalf of another firm, the two firms enter into a relationship of principal and agent. For example, pharmaceutical companies often subcontract clinical trials to specialized firms known as contract research organizations (CROs), which capture almost one-fifth of the overall clinical development budget in the US (Azoulay 2004). Similarly, tasks of R&D in automobile engineering are frequently subcontracted to specialized firms (Smitka 1991). As the party carrying out R&D on the ground, the agent may learn information about the evolving technology much faster than its principal (Teece 1980; Acharya et al. 2011). As a result, every time new laboratory or market conditions shift the technology outside the contract's scope, there is an *opportunity* for the agent to exploit the principal, for example by making less efforts, using cheaper but less suitable inputs, or renegotiating constantly for better terms (such as higher prices or later delivery). The agent's informational advantage, in other words, allows it to behave *opportunistically* (Gershkov and Modovanu 2009; Alonso et al. 2008). Through opportunistic shirking or renegotiation of contracts, the agent can in effect offload its own market or technological risks onto the principal instead. How may the principal smooth such risks posed by opportunistic agents? Should the principal exert hierarchical *control* over the agent (by taking over ownership through vertical integration) or grant *discretion* to the agent as an independent contractor? Between liberal market and strategically coordinated economies, whose technological innovators are "subordinates," and whose innovators are "independent contractors"? In technological innovation, how do liberal market and strategically coordinated capitalism differ in their choice between *control* and *discretion* as two different strategies of managing opportunistic volatility? These are the central questions of this chapter.

My focus on volatilities driven by the *opportunistic firm* also sets the book apart from studies that focus on volatilities caused by conditions *exogenous to the firm*, such as various macroeconomic conditions, government policies, and the international trade environment (Keefer and Stasavage 2003; Kim 2007; Kayser 2005). An individual firm may of course *adjust* strategically to these macro conditions, but the firm itself rarely has enough power to *change* or *eliminate* these conditions. For instance, Katzenstein's (1984, 1985) "domestic compensation" thesis is a good example of how small open economies *adjust* to volatilities imposed by the world economy. On the other hand, a firm as the principal may in fact *eliminate*, rather than merely adjust to, the risks of its agent behaving opportunistically, for example by taking over ownership and control, in effect establishing a relationship of hierarchy (Milgrom and Holmström 1994; Hart 1995; Schmitz 2006). In other words, how firms manage the relational side of their activities (the issue at the core of the varieties of capitalism literature) may have a particularly clear effect on *opportunistic* volatility.

This focus on opportunistic volatility builds a connection with Chapter 2's discussion of how different types of capitalism manage relationship risks: as I argued in that chapter's setting of financial markets, firms in liberal market capitalism *pool* risks *across the market* by releasing public information, while firms in strategically coordinated capitalism *smooth* risks *over time* by sending insider information as a signal of commitment. The current chapter's setting of principal–agent relationship in innovation provides a very different environment for risk management than Chapter 2's setting of firm–investor relationship in financial markets. Stock market financing is an important tool for risk pooling. There, a single firm may be financed by potentially thousands of individual investors, who agree to disagree over their own beliefs about the firm's prospect. The mutual offsetting of pessimistic and optimistic beliefs across the market helps the firm reduce the possibility of an all-out investor flight (Harris and Raviv 1993; Harrison and Kreps 1978). By contrast, when one firm pays another firm to supply new technologies, the margin for such cross-sectional risk pooling is far more limited. First, the principal cannot feasibly contract simultaneously with all suppliers in the market. Second, because innovation has sunk costs (Stiglitz et al. 1987), even if the principal is dissatisfied, it may balk at immediate contract termination, and choose only not to renew the contract the next time. In other words, the process of R&D necessarily binds the principal to a fixed relationship with the agent, forcing it to smooth risks over the period of that relationship. But what kind of relationship? Is it a relationship where the principal controls the agent as a subordinate, or leaves discretion to the agent as an independent contractor? In other words, instead of comparing risk pooling with smoothing, here I examine how different types of capitalism choose between two different strategies of risk smoothing: *control* versus *discretion*. In this chapter, I argue that firms in liberal market capitalism prefer *control* over the agent, while firms in strategically coordinated capitalism grant *discretion* to the agent.

In practice, how do firms exercise such *control* or grant such *discretion*? Starting with Ronald Coase (1937, 1960), scholars of property rights have long pointed out that firms use ownership to delineate the boundary of their control (Arrow 1974; Demsetz 1988; Hart 1995; Milgrom and Holmström 1994; Barzel 1997; Tandelis 2002). If the principal firm acquires the agent firm as a subdivision through vertical integration, the agent's fixed assets as well as production output become the property of the principal. Since the agent's R&D work requires the use of these fixed assets, property rights give the principal not only the authority to determine the agent's action in situations unanticipated by the contract, but also greater bargaining power if the agent attempts to renegotiate for better terms. By contrast, if the agent is an independent contractor, it has full discretion over its own action in circumstances outside the contract's scope, and if the relationship with the principal is dissolved, it has the right to dispose of its own output, which enhances its bargaining power (Hart 1995; Schmitz 2006; Lerner and Malmendier 2010).

In strategically coordinated capitalism, there are various dense institutions that facilitate long-term relational contracting, such as encompassing and cohesive business federations, patient capital, and political endorsement by the state (all captured explicitly in the Hicks–Kenworthy and Hall–Gingerich indices of coordination used throughout the book). With relational contracting, the principal offers to renew the relationship with the agent repeatedly as long as the agent cooperates and delivers satisfactory performance. This prospect of long-term relationship discourages the agent from short-term opportunism even if the agent enjoys full discretion as an independent contractor (Chassang 2010; Lafontaine and Raynaud 2002). In other words, in economies with deep relational institutions, the principal can reduce opportunism even if it has no ownership *control* over its agent. Liberal market capitalism, by contrast, lacks institutions that facilitate long-term cooperation between firms. Without the prospect of long-term relational commitment, the principal has to use *control* under the ownership hierarchy to prevent agent opportunism (Lafontaine and Raynaud 2002; Klein 2002). In other words, different types of capitalism adopt different strategies to manage the principal–agent relationship during technological innovation. In liberal market economies, firms prefer *control*, and obtain technologies from divisions under their own ownership hierarchy; in strategically coordinated economies, firms prefer *discretion* by the agent, and obtain technologies from independent contractors.

This finding joins many scholars of comparative political economy in highlighting an interesting irony about liberal market economies: despite their reputation for more fluid market institutions, these economies are precisely where players rely most heavily on hierarchical control to govern economic relationships (Piore and Sabel 1984; Chandler 1977, 1990; Lazonick 1990; Hollingsworth and Boyer 1997; Djelic 1998; Whitley 1999; Fligstein 2001). This chapter identifies a particularly counterintuitive type of control: although the exercise of hierarchical control is commonly understood to rely on some ability to *monitor* actions and *enforce* contracts (as scholars on the employment hierarchy and state intervention point out (Schmidt 1990; Vail 2009; Levy 2006; Hancké et al. 2007; Howell 2005; Shapiro and Stiglitz 1984)), I identify a type of "rule by fiat" that thrives precisely where the actions of the ruled *cannot* be monitored or enforced by contract. Governing actions that are *left out of* formal contracts, such control is *residual*, based on the *ownership* of financial and physical capital. As I will show, such control is not only subtle (without forcing the ruled to divulge their private information) but also efficient (it binds only when the relationship becomes more volatile).

Adding a further twist to this surprising form of control, I show that, despite its elegant power, control turns out to be *worse than having no control at all*. While *control* through the ownership hierarchy may reduce opportunistic volatility, it may also inadvertently reduce innovation output, by downgrading the agent's motivation (Milgrom and Holmström 1994; Holmström 1999). As a large theoretical and experimental literature has documented extensively (Williamson 1985; Demsetz 1988; Falk and Kosfeld 2006; Bartling et al. 2012), *control* distorts not only the agent's incentives but also its beliefs. When the agent does not own the input or output of its own technological production and cannot share residual profits, the agent's incentives to raise output are "low-powered." Furthermore, the agent infers from its own lack of *discretion* that the principal anticipates the agent to have little motivation, which in turn discourages the agent from reciprocating with high efforts, completing a "self-fulfilling prophesy of distrust" (Luhmann 1968). By contrast, if the principal leaves *discretion* to the agent as an independent contractor, these "hidden costs of control" (Falk and Kosfeld 2006) are eliminated. Therefore, in innovation, although both *control* (under the ownership hierarchy) and *discretion* (under relational contracting) may reduce opportunistic volatility, the former also carries the cost of lower innovation output. In other words, liberal market economies face a tradeoff between the need to lower volatility and the need to raise innovation output. The more one goal is achieved, the harder it is to achieve the other. I refer to this effect of liberal market economies as the *demotivation effect*, whereby the use of control crowds out agents' motivation to raise output. In strategically coordinated economies, by contrast, the twin goals of high output and low volatility do not defeat each other, and as I will show empirically, they may in fact be complementary.

In the first main section of this chapter, I draw from the literature on property rights and industrial organization to develop a theory of *control* (via the firm hierarchy) and *discretion* (via relational contracting) as different innovation strategies chosen by different types of capitalism. Based on multiple years of firm survey data, I present empirical evidence that firms in liberal market capitalism emphasize *control*, and prefer R&D carried out by their own subordinates, while firms in strategically coordinated capitalism emphasize *discretion*, and prefer R&D from independent contractors.

On this basis, in the second main section of the chapter I examine the tradeoff between volatility and innovation output for different varieties of capitalism. Because firms in liberal market economies smooth risks through *control*, they have to reduce opportunistic volatility at the cost of lower innovation output. By contrast, firms in strategically coordinated economies leave *discretion* to their R&D agents, and as a result they face no dilemma between low volatility and high output. Using a combination of firm survey data on innovation output and OECD data on firm revenue volatility, I show that while reductions in volatility have a notably negative impact on innovation output in liberal market economies (i.e. the *demotivation effect*), this volatility–output tradeoff vanishes in strategically coordinated capitalism. In fact, in very strongly coordinated economies, low volatility and high innovation output are complements: reductions in volatility have a real effect in *raising* innovation performance.

Who Controls and Who Leaves Discretion during Innovation?

Technological innovation is a stochastic process with constant arrivals of unanticipated information and states of nature. As a result, when an agent firm carries out R&D on behalf of a principal firm who delegates the tasks and collects the output, the agent often learns "on-the-ground" information about the innovation's evolution and its market implications more quickly and fully than the principal. On one hand, the agent is several steps ahead of the principal in learning new information; on the other hand, the evolving nature of new technologies requires the agent to respond timely, rather than waiting until the principal itself fully observes the new information (Simon 1951; MacLeod 2002; Gershkov and Modovanu 2009). In other words, the agent may take advantage of the principal's informational lag and behave opportunistically in the relationship. There are at least three channels through which the agent may leverage its own informational advantage opportunistically.

First, it may mislead the principal about whether the innovation process has moved to a situation unspecified in the contract (Acharya et al. 2011). No matter how elaborate, no contract between the principal and the agent can fully anticipate all the contingencies that await the R&D process, and as a result, no contract can fully specify how the agent should behave under these circumstances (Tirole 2009). Each arrival of such new circumstance, therefore, is an opportunity for the agent to deviate from the principal's preferred action. For example, in order to win a drugs patent race, a principal firm may hire a more specialized agent firm to deliver a new lead compound. If there is a significant change in the thermal property of the compound during its R&D, the agent may advise that the new thermal environment for the compound's development is significantly different from the one governed by the contract, and therefore request to switch to a less difficult but slower method. This, of course, is bad news to the principal in the patent race. The principal may disagree with this switch to the slower method, and engage in costly efforts (such as third-party expert verification or contract disputes) (Barzel 2002) to counter that the change in thermal environment is too marginal to exceed the contract specification. These costly efforts, however, are wasteful if the agent had been truthful in its advice. Alternatively, the principal can defer to the agent's advice and accept the switch to the slower method. This, however, harms the principal if the agent had in reality exaggerated the change in the thermal environment. The principal's informational problem is that the agent knows much better whether its own advice is truthful or not. Although later the principal may gradually glean additional information about the new chemical's true thermal properties (for example, through repeated use) (Nelson 1970; Darby and Karni 1973) and therefore learn whether the agent had been truthful in its advice, this full revelation may only occur well after the agent's contract has reached completion.

Second, even if both sides agree that the innovation process has moved outside the scope of the contract, the agent may still mislead about how it had exercised its choices where the contract does not govern (Aghion et al. 2013; Kamenica and Gentzkow 2011). For example, both sides may agree that a sudden change in market conditions unanticipated in the contract gives the agent the freedom to find a new supplier for the raw materials used in its R&D. The agent firm can choose either a reputable or untried supplier. The principal prefers the former because it is better for the innovation project but the agent prefers the latter because this new supplier entices the agent with private rents (such as discounts on raw materials for the agent's side projects unrelated to the principal) (Lerner and Malmendier 2010). The agent may proceed with the new supplier and advise that the negative impact on the principal's R&D is minimal. For its part, the principal may invest in costly adjustments in production and marketing to offset the negative impact on its R&D. These costly adjustments, however, are wasteful if the agent had been truthful in its advice (Gershkov and Modovanu 2009). Alternatively, the principal may choose not to make adjustments. This, however, harms the principal if the agent had downplayed the impact of using a less reputable supplier. Again, the principal's problem is that the agent knows much better whether its own advice is truthful or not. Of course, the agent's informational advantage does not have to stem from its technological expertise. The agent may also have private information about the upstream market that it deals with, for example, and mislead the principal about the true extent to which a change in supplier is necessary. Along the same vein, Azoulay (2004) documented how CROs often engage in opportunistic "bait and switch," and give customer firms "rookie" researchers instead of "A-teams."

Third, even when the R&D process is complete and the new technology created, the agent firm can still mislead the principal firm and renegotiate to its own advantage. For example, the agent firm may advise that the new technology's properties and demand in downstream markets have far exceeded the range anticipated in the original contract, and therefore renegotiate a higher price from the principal. In the event that the renegotiation was unsuccessful and the contract was dissolved, the agent firm retains the technology it has created and the principal firm gets back the payment it has made to the agent (Hart 2002; Schmitz 2006). If the agent had been truthful in its opinion about the new technology, the principal may be better off agreeing to a higher price than letting the contract dissolve and the agent pass the technology to the principal's competitors. By contrast, if the agent had exaggerated the product's potential, then the principal may be better off refusing to renegotiate, knowing that, by retaining the technology, the agent's outside option in the market is limited. However, again the principal's informational problem is that the agent may know far better than the principal about the true properties of the technology. Since new technologies are "credence goods" whose real properties are only revealed slowly (and imperfectly so) after repeated use (Darby and Karni 1973), the principal may only learn about them long after the current contract renegotiation has been completed.

In short, technical and market volatilities inherent to the innovation process present many opportunities for the agent to exploit the principal's relative disadvantage in information, and extract more surplus from the principal. In effect, through such opportunistic behavior, the information-rich agent passes its own market or technological risks onto the shoulder of the information-poor principal. Is it possible for the principal to smooth such opportunistic risks over the course of the R&D relationship? In strategically coordinated capitalism, there are various institutional resources that facilitate long-term relational contracting, such as encompassing and cohesive business organizations with the capacity to coordinate interfirm relationships (Crouch 1993; Sako 1992; Coleman and Grant 1988; Streeck 2005), deep ties to large-stake and patient investors who prefer "voice" to "exit" (Hicks and Kenworthy 1998; Gourevitch and Shinn 2005; Culpepper 2011), and cooperative interaction between cohesive government agencies and centralized business/labor organizations (Ornston 2012; Busch 1999; Levy 2006; Martin and Swank 2008). In strategically coordinated capitalism, the rich presence of these institutional resources helps firms smooth opportunistic risks by establishing long-term cooperative relationships: the principal continues to extend the relationship (e.g. renew the contract or be a repeat customer) as long as the agent cooperates and delivers satisfactory performance. This prospect of long-term relationship discourages the agent from short-term opportunism even if the agent enjoys full discretion as an independent contractor (Chassang 2010; Lafontaine and Raynaud 2002; Klein et al. 1996). In other words, in economies with deep relational institutions, the principal can reduce opportunism even if it has no ownership *control* over its agent.

Liberal market economies, by contrast, lack such strategic coordination institutions to lengthen firms' horizons or facilitate long-term relational contracting. When there are no committed patient investors, deep interfirm networks, or political endorsement by the state, it is harder for the principal to credibly commit to renew the relationship continuously, and as a result the agent is no longer motivated to refrain from opportunistic behavior. Similarly, when the lack of coordination institutions shortens the agent's relational horizon, the principal no longer has the relational means to deter agent opportunism. During the fluid and complex R&D process, the agent on the ground learns new information more quickly than the principal, and needs to act before the principal catches up in information (Acharya et al. 2011; Simon 1951; MacLeod 2002). Even if the principal may later obtain additional information and gradually learn whether the agent had acted opportunistically, the principal has limited room in responding to this revelation of opportunism: the constantly evolving R&D would have moved on, and the contract may have long reached its completion. The best the principal can do is to choose not to renew the contract, but this cannot motivate agents who have *short* relationship horizons.

In other words, when *markets* for new technologies are not embedded in dense institutions of strategic *coordination*, one may naturally expect firms to be more volatile in their relationships and face higher opportunistic risks. This comparison between *markets* and strategic *coordination* is of course at the very heart of the literature on varieties of capitalism (Hall and Soskice 2001; Martin and Swank 2012; Culpepper 2011; Amable 2003; Gourevitch and Shinn 2005; Iversen and Soskice 2009; Thelen 2004; Ornston 2012; Lazonick 2007). This analytical angle, for example, is captured vividly in two of the most influential terms used in comparing capitalism: "liberal market economies" and "coordinated market economies" (Hall and Soskice 2001). Nevertheless, as Hall and Soskice (2001, pp. 8–9) pointed out, strategic *coordination* should be compared not only with *markets*, but also *hierarchies*. While both *markets* and strategic *coordination* rely on *negotiation* to coordinate behavior, *hierarchies* also rely on *fiat* to control behavior. As a result, the nature of relationship governance under the ownership or employment hierarchy is fundamentally different from governance by markets or strategic coordination (Fligstein 2001; Hollingsworth 1993; Farrell 2009; Best 1990). In other words, when hierarchies are joined to the studies of markets and nonmarket coordination, the comparison of different capitalisms may yield considerably richer insights. When does a market or nonmarket coordination relationship between firms become a relationship ordered by hierarchy? Inside the hierarchy, when is the time for "action by consensus" and when is the time for "rule by fiat"?

The extensive literature on property rights, contracts, and industrial organization has generated considerable insights on these questions (Arrow 1974; Williamson 1985; Grossman and Hart 1986; Demsetz 1988; Milgrom and Holmström 1994; Barzel 1997; Tandelis 2002). As Ronald Coase (1937, 1960) points out, a relationship between the principal and the agent as equal peers becomes hierarchical when the principal firm owns the intermediate capital (such as physical or financial assets) which the agent uses to make the product. This transforms the agent from an independent contractor into a subdivision (or employee) of the principal firm, who in turn becomes the residual owner. Instead of *buying* from *another* firm, the principal is now in effect *making* the product from *its own* subdivision. In other words, the allocation of property rights in the production process sets the *boundary of the firm*, i.e. where one firm ends and another begins (Williamson 1985). Even more importantly, the allocation of property rights also demarcates the principal's *boundary of control*, because residual ownership gives the principal the exclusive right to decide how the productive assets should be used in circumstances unspecified in the contract, and how the output from these assets should be disposed of. In other words, residual ownership leads to residual control, that is, control in circumstances left out of the contract and in disposing the final output (Grossman and Hart 1986; Hart 1995; Barzel 1997; MacLeod 2002).

Whenever one party contracts another to produce a product/service, it is often either too costly or simply infeasible to specify all possible contingencies during the production process (Tirole 2009). If the principal owns the physical or financial assets used in the production, then the agent has no right to unilaterally determine the use of these assets in circumstances outside the contract's provision. For example, a trucker hired to drive the truck owned by a shipping company has no right to drive the vehicle outside designated routes, or physically modify the truck (for example, altering ladder frames in the chassis) to adjust to the goods transported, unless this option is specified in the contract (Hart 2002). Similarly, a firm's subdivision assigned to making parts cannot use the firm's physical and financial capital to make other products, or sell off or convert the parts made, without managerial permission from above. Therefore, by exercising residual control over the production capital used by the agent, the principal has the exclusive right to determine how the agent should behave where the contract is incomplete. In a nutshell, in answer to the questions posed two paragraphs earlier, a relationship becomes ordered by hierarchy *when the principal "makes" rather than "buys"*; inside the hierarchy, the principal starts to "rule by fiat" *when circumstances unanticipated in the contract arise.*

As the neocorporatist wage bargaining literature pointed out, even within the firm hierarchy, there is significant room for bargaining. In this bargaining environment, residual ownership also gives the principal stronger bargaining power, because if the contract is dissolved the principal has the right to dispose of the agent's output in any way it sees fit. Going back to the earlier example of the subdivision making parts for the firm, after the parts are made, they become the property of the firm. This puts the firm in a strong bargaining position if employees from the subdivision decide to renegotiate for higher compensation: in the event the renegotiation fails and the workers leave, they cannot take the parts with them. As a result, the subdivision's outside option (and bargaining power) is limited (Schmitz 2006; Lerner and Malmendier 2010).

By contrast, if the agent rather than the principal owns the intermediate capital used in the production, then the agent has discretion over how to use the capital in situations ungoverned by the contract and how to dispose of the output made from the capital if the contract is dissolved. For example, by owning his own truck, the truck driver in effect becomes a *trucking company* contracted by the shipping company for its service, and he is free to alter his own truck's ladder frame to accommodate the goods transported, even if the contract does not state whether this is permissible (Hart 2002). Similarly, if the part supplier actually owns the physical and financial capital used in making parts for the customer firm, this turns the supplier from a subdivision of the customer firm into an independent contractor. As a result, the supplier is now free to use its own assets for purposes other than making parts for the customer firm, and in the event that the contract is dissolved, the supplier has the right to sell off the parts to other buyers or convert the parts in any way it sees fit.

In a nutshell, whoever has residual ownership also has greater control when the principal–agent relationship moves to a situation unforeseen by the contract. For this reason, although firms in liberal market economies cannot effectively discourage opportunism through long-term relational commitment, they can suppress opportunism by maintaining residual ownership, which allows them to exercise residual *control*. In fact, residual ownership may allow the principal to shut down all three channels of agent opportunism outlined at the beginning of this section.

First, even if private information enables the agent to mislead the principal about whether the innovation process has moved to a circumstance unspecified by the contract (Acharya et al. 2011), this does not give the agent the opportunity to deviate from the principal's preferred action. When a principal firm in a drugs patent race asks an independent contractor to develop a chemical compound, the agent may advise that the compound's discovery has entered a thermal environment not governed by the contract, and hence switch to a slower method, which harms the principal in the drugs race. The principal cannot enforce the agent's behavior under this circumstance because the agent claims it is outside the contract's specification. By contrast, if the agent is merely a "development division" of the principal firm, the physical and financial capital used in the R&D work is owned by the principal firm. Residual ownership gives the principal the right to unilaterally determine how to use these assets where the contract does not specify. As a result, even if the agent (falsely) claims that the drug development has entered an environment not governed by the contract, this does not give it the right to redeploy the company's resources to the slower production method without the company's permission (Van den Steen 2010). In other words, even if the agent's better information may allow it to mislead about whether the R&D has moved to a "new state" unanticipated in the contract, the principal's residual control shuts down this "new state" as a venue for the agent to behave opportunistically.

Second, *control* can also prevent opportunism when the agent's private information is not about *the state of* the innovation, but about its *own choice of action* in this state (Aghion et al. 2013; Kamenica and Gentzkow 2011). For example, both sides may agree that sudden changes in market conditions require the agent to search for a new supplier of raw materials used in its R&D work, a situation unanticipated in the contract. Since how the agent should conduct the search is not specified in the contract, if the agent is an independent contractor who uses its own financial capital to procure raw materials, it has the freedom to choose a less reputable supplier to collect private rents (such as private discounts) useless to the principal. If it then misleads the principal about the extent to which this supplier will have a negative impact on the R&D, the principal will not be able to adjust its resources to this impact effectively (Gershkov and Modovanu 2009; Lerner and Malmendier 2010). By contrast, if the principal owns the agent as a subdivision, the principal has the exclusive right to determine how its own financial capital should be deployed. In other words, the principal has the unilateral authority to select a more reputable supplier in order to safeguard the integrity of the R&D process. Again, by taking away the agent's choice of action, the principal's residual control prevents new market conditions from becoming a channel for the agent to behave opportunistically.

Third, even where the principal relies on negotiation rather than "rule by fiat," residual ownership may give the principal stronger bargaining power, which prevents the agent from exploiting its own informational advantage to renegotiate for better terms (Schmitz 2006; Tirole 2009). For example, when the agent's R&D task is complete and the new technology created, it may advise that the new technology's quality has far exceeded the standard anticipated in the original contract, and therefore renegotiate a higher price from the principal. If the agent is an independent contractor who uses its own physical and financial assets to conduct the R&D, it retains ownership of the new technology in the event that the renegotiation fails and the contract is dissolved. Based on its own superior information, the agent may exaggerate its claim about the potential market demand for this technology, which it can retain and sell elsewhere if the principal refuses to renegotiate. Armed with this better "outside option," the agent can weaken the bargaining position of the principal, who will be under greater pressure to concede and prevent the technology from being passed to competitors. By contrast, if the principal firm owns the agent as a "research division," after the new technology is created, it becomes the property of the firm. This puts the principal in a strong bargaining position if employees from the subdivision decide to renegotiate for higher compensation: in the event the renegotiation fails and the employees leave, the technology remains with the firm (Hart 1995, 2002). As a result, even if the agent may exaggerate the quality of the technology, the principal's residual ownership over the technology limits the agent's ability to exploit misinformation and renegotiate for better terms.

Therefore, although liberal market economies cannot discourage opportunism through long-term relational commitment, they may suppress opportunism using *control* under the ownership hierarchy. When the principal chooses to have its R&D done by its own subdivision rather than an independent firm, it brings the agent under its ownership hierarchy. This residual ownership not only allows the principal to "rule by fiat" but also enables it to negotiate with greater bargaining power, both of which limit the agent's ability to opportunistically exploit its own informational advantage.

Scholars have long pointed out that, despite its name, "liberal market" capitalism often relies heavily on "control and sanctions" in reality. This is not only reflected in how employers approach their relationship with employees (Kenney and Florida 1993; Piore and Sabel 1984; Lazonick 1990; Hollingsworth and Boyer 1997; Whitley 1999; Fligstein 2001) but also in how the state approaches its relationship with businesses and labor (Moran 2006; Howell 2005; Ginsburg 2005). While this chapter's argument that liberal market economies rely more on residual control is firmly consistent with this central theme, it also has some subtle differences from existing accounts. Because existing accounts examine the use of control mostly from the perspective of *employers* or the *state*, there is already a *hierarchy inherent to* the relationship being studied: employers inherently have a right to "discipline" workers, and the state inherently has a right to "regulate" businesses or labor. This chapter, by contrast, examines an inherently *lateral* relationship from the perspective of *firms*: how do they deal with each other when they sell each other technologies? Unlike the employment or state hierarchy, "cutting loose" by giving the other side complete discretion is a feasible (and as will be shown later, very rewarding) option when firms contract with each other. This makes the choice between *control* and *discretion* less trivial, the decision to choose *control* more puzzling, and the nature of such *control* more interesting. The type of ownership-based control identified in this chapter is particularly interesting in two ways. First, it is subtle: it prevents others from opportunistically exploiting hidden information *without* forcing them to give up the information. Second, it is efficient: it binds *only when* the relationship is most volatile.

When the principal owns the physical and financial capital used in the agent's R&D work, residual ownership gives the principal the authority to "rule by fiat" in new circumstances ungoverned by the contract. As a result, even if the agent possesses hidden information, *control* removes the agent's discretion to exploit such information precisely where it is most likely to be exploited opportunistically. Of course, besides removing the agent's *discretion*, residual ownership may also reduce the principal's informational disadvantage through an even blunter channel, by directly removing the agent's *hidden information*. With the agent under its own hierarchy, the principal has greater capacity to extract the agent's hidden information, either through closer monitoring (Aghion et al. 2013) or stronger disclosure, for example ordering the agent as a subdivision to release hidden information "by fiat." Instead of bluntly *removing information*, this chapter's theory of control highlights a more subtle mechanism of *removing the discretion to act on information*. This puts in sharper relief the power of residual control: opportunistic exploitation of hidden information may be eliminated, even if hidden information is *not* revealed.

Furthermore, monitoring, verification, and enforcement are traditionally understood to be indispensible elements of control. For instance, employers cannot effectively discipline workers, and the state cannot effectively regulate businesses, if the actions of workers or businesses cannot be observed. Even in more "lateral" relationships such as contracting between two firms, the use of legal "rule by fiat" (to obtain timely payment or goods delivery) relies on the ability to monitor and verify the other side's actions. In other words, control, somewhat tautologically, can only work when the controller already has a "good grasp" of the situation. This tautology underlines a source of inefficiency in traditional forms of control: the exercise of such control is more effective precisely when there is less uncertainty about the relationship, and hence, less need for the very exercise of control. This chapter, by contrast, identifies a type of "rule by fiat" that thrives precisely when the actions of the ruled *cannot* be monitored or enforced within the contract, in other words, when the relationship is *most* volatile. Because the R&D process is not only stochastic but also rapidly evolving, the principal firm and its agent in technological innovation have to face continuous arrivals of new circumstances unanticipated by their contract. On one hand, since they are impossible to monitor or enforce, these outside-contract circumstances present the principal with the greatest relationship volatility; on the other hand, "rule by fiat" under the ownership hierarchy kicks in precisely when circumstances move outside the contract range (Simon 1951; Williamson 1985; Grossman and Hart 1986). In other words, residual ownership is a very efficient instrument of control: it binds *only when* the relationship is most volatile. Therefore, in liberal market economies, where agent opportunism cannot be effectively discouraged with long-term relational commitment, residual control becomes an important alternative means of suppressing opportunism.

Overall, the discussion above suggests that, depending on the strength of coordination, different types of capitalism will adopt different strategies to manage the principal–agent relationship during technological innovation. In strategically coordinated capitalism, the principal can use the prospect of long-term relational commitment to discourage its agents from short-term opportunism. In other words, the principal can reduce opportunistic volatility without taking away agents' *discretion* as independent firms. In liberal market capitalism, by contrast, firms lack the institutional resources to offer long-term relational commitment. As a result, they have to *control* their agents under the ownership hierarchy in order to suppress opportunistic volatility. In other words, while firms in liberal market economies prefer *control*, and obtain technologies from divisions under their own ownership hierarchy, firms in strategically coordinated economies grant *discretion*, and obtain R&D from independent contractors. The more coordinated the economy, the less firms resort to internal ownership *control*, and the more they allow external contractors to conduct R&D on their behalf.

> H1: the more coordinated the economy, the less firms spend on internal R&D by subordinates relative to external R&D by independent contractors.

Empirical Analysis

The testing of hypothesis H1 across varieties of capitalism requires data that is not only comparable across a wide range of OECD countries but also conceptually capable of distinguishing whether firms' innovation efforts are based on internal R&D by subordinates or external R&D by independent contractors. More specifically, when a firm spends on R&D, is the spending used to support research carried out within its own boundary, or acquire research carried out by independent firms outside its boundary? These two types of R&D spending are defined respectively as "intramural" and "extramural" R&D by the *Oslo Manual* (OECD 2005), the internationally adopted methodological guide for measuring innovation. For data on these two types of R&D, I again turn to CIS, jointly administered by Eurostat and Directorate-General Enterprise of the European Union. There are several benefits from using CIS data for the current analysis. First, as an enterprise survey, it asks firms directly about the nature and purpose of their R&D activities; second, its questionnaire relies directly on the methodological guidelines from the earlier-mentioned *Oslo Manual* in describing all aspects of innovation; thirdly, it is carried out on a common, standardized, questionnaire across a large number of countries, and the survey's subject pool encompasses all enterprises with ten or more employees. Although the actual survey is carried out on the level of individual enterprises, its results are aggregated by Eurostat to the sector level before being made available to the public, in order to maintain confidentiality for the firms polled. Without direct firm-level observations, the researcher has to construct measures that are sector-level correlates of firms' choices between intramural and extramural R&D.

Starting from the third survey wave in 2001, each wave of CIS provides two variables that measure each sector's total enterprise spending on respectively extramural and intramural R&D as a percentage of the sector's total turnover. To capture the relative extent to which a sector's firms spend on intramural versus extramural innovation, I use the ratio of intramural R&D spending over extramural R&D spending as the dependent variable for H1. In total, there are four waves (the third wave in 2001, fourth wave in 2004, fifth wave in 2006, and sixth wave in 2008) that provide sufficient data for the variables to be included in the analysis. However, because these are not panel surveys, and because there are very few years of observation (four survey waves) relative to three times as many cross-sectional units, I analyze data from each survey as a separate cross-sectional analysis and use the findings from different surveys as robustness checks against each other.

Within the tight constraint of available survey questions in CIS, I also include some control variables that may further affect whether firms choose to carry out R&D extramurally or intramurally. Clearly, agent opportunism is not the only factor that determines the principal's choice between "making" innovation intramurally and "buying" innovation extramurally. If an extramural agent (i.e. an independent contractor) possesses niche experience and proficiency in the R&D task, then agent *expertise* will attract the principal, which potentially offsets the effect of agent *opportunism* in driving the principal away towards intramural innovation. Although there is no data in CIS that directly measures extramural agents' expertise, each wave of the survey does provide measures that indirectly tap into some mechanisms that allow a principal firm to reduce its own reliance on the expertise of extramurally contracted agents, which may in turn encourage more intramural R&D. Two variables in CIS report the percentage of firms in a given sector that cited "peer-based" third-party platforms as important sources of technical information for their own innovation, focusing respectively on "conferences, trade fairs, exhibitions" (data available for all four surveys) and "scientific journals and trade/technical publications" (data available from the 2004 survey onward). Along the similar vein, three other variables in CIS provide the percentage of firms citing "organization-based" third-party platforms as important sources of technical information for their own innovation, focusing respectively on "government or public research institutes" (data available for all four surveys), "universities or other higher education institutes" (data available for all four surveys), and "professional and industry associations" (data available from the 2004 wave onward). The more the principal can exploit these platforms as sources of learning for its own innovation, the less its need to contract the R&D task to an extramural agent. While firms' ability to learn innovation through a third party may *discourage* extramural R&D by reducing their need for extramural agent expertise, learning through a third party may also have the opposite effect of *encouraging* extramural R&D through a different channel: as the principal becomes better informed about the innovation, there is less room for agent opportunism even if the R&D is contracted out. Between these two mutually offsetting effects of learning through third parties, which one dominates will be an empirical question, to be settled in the process of CIS data analysis.

Another potential incentive for firms to carry out their own R&D rather than subcontracting is the need to protect the innovation output through secrecy. However, secrecy is not always an effective measure to protect new technologies. If new technologies are patented, the information disclosure requirements of patent application will prevent the firm from effectively concealing the new technological knowledge. As an alternative, firms may capitalize on complex cumulative system technologies and engage in mutual licensing, with several firms developing complementary patents which have to be utilized in combination in order to complete the entire system. Collectively these complementary innovations are defined as a "thicket": ownership over individual components in the thicket can be legally allocated to separate firms but their economic value cannot be realized without being assembled into one system (Reitzig 2004). Each firm in the thicket therefore uses its own innovation as a bargaining chip in acquiring complementary innovation extramurally from others. In other words, when firms protect their own innovation by embedding it in a complex system, they become both innovation suppliers and customers to one another in the same thicket, which turns more R&D into extramural activities (Merges and Nelson 1994; Arora and Fosfuri 2002). In the 2001 wave of CIS, two variables report the percentage of firms in a given sector that cited, respectively, secrecy and complex system design as important methods of protection for their innovation output. I include both as controls. While secrecy should drive firms towards more intramural R&D, complex system design should lead to more extramural R&D. These two variables, unfortunately, are no longer available after the 2001 survey wave, so they are only included in the analysis of the 2001 wave.

Similar to the analysis of CIS data in Chapter 3, I again use the SUR (Zellner 1962) estimator. As mentioned before, this procedure directly accommodates the presence of interdependent errors across equations, and therefore can simultaneously estimate multiple equations with (nearly) identical variables and a common set of data. As noted in Chapter 3, this SUR technique of analyzing survey-based innovation data follows other works in the innovation literature, such as Betts (1997) and Piva et al. (2005).

Who Faces a Dilemma between Volatility and Output?

In order to reinforce the robustness of findings, I analyze the survey data using alternatively the Hicks–Kenworthy and Hall–Gingerich measure of coordination. With two equations for each SUR estimation, four surveys, and two different measures of coordination, this exercise produces in total sixteen sets of findings that may serve as robustness checks against each other. In order to ensure concise presentation while still preserving a flavor of the findings, the eight columns in Table 5.1 alternate between the two measures of coordination, using Hicks–Kenworthy for even-number ordered surveys (third and fifth waves) and Hall–Gingerich for odd-number ordered surveys (fourth and sixth waves).

I start the discussion with the 2001 survey (first two columns). The dependent variable, to reiterate, is the relative extent to which firms engage in intramural versus extramural innovation, measured as the ratio of intramural R&D spending over extramural R&D spending. Across countries in this analysis, the average ratio of intramural over extramural R&D is 1.28. As the coefficients for coordination indicate, a one-unit increase in coordination will reduce the ratio of intramural over extramural R&D spending by slightly more than 0.5. This reduction in the relative proportion of intramural R&D is equal to about 42 percent of the average extent to which firms engage in intramural relative to extramural R&D across the countries.

This implies that each standard deviation increase in coordination will tilt the relative balance away from intramural R&D by 0.16 on the ratio, equal to more than a tenth (12 percent) of the average extent to which firms engage in intramural relative to extramural innovation. Compared with minimum coordination, an economy with average coordination (0.51 on the Hicks–Kenworthy index) will reduce, by almost a fifth of the average, the extent to which firms conduct their own R&D rather than contracting it out to other firms. With all control variables at their mean, firms in a country with median strength of coordination split just under half and half (0.89 on the ratio) between carrying out their own R&D and contracting it out to others. A standard deviation more coordination will slash the ratio to 0.73, while two standard deviations will cut the amount of intramural innovation to barely more than half of extramural innovation (0.56 on the ratio). Moving in the other direction, by contrast, two standard deviations below average coordination will result in more intramural than extramural innovation, by the ratio of 1.21.

To gain some further substantive meaning from the results, I use the point estimates to derive the difference between specific countries in the extent to which their firms choose "make" or "buy" in innovation. For example, if American firms (0.02 on the Hicks–Kenworthy index) are endowed with coordination institutions of German strength (0.80 on the index), their ratio of intramural over extramural innovation will decline by 0.41. In other words,

Table 5.1. Ratio of Intramural over Extramural R&D, Twelve OECD Countries

Community Innovation Survey Wave	Third Wave (2001) Equation a	Third Wave (2001) Equation b	Fourth Wave (2004) Equation a	Fourth Wave (2004) Equation b	Fifth Wave (2006) Equation a	Fifth Wave (2006) Equation b	Sixth Wave (2008) Equation a	Sixth Wave (2008) Equation b
Coordination	−0.54(0.11)***	−0.62(0.25)**	−0.31(0.12)**	−0.26(0.10)**	−0.69(0.34)	−0.77(0.35)*	−0.40(0.15)**	−0.48(0.15)***
Conferences, fairs, exhibitions as important source of information for innovation	0.17(0.04)***	0.16(0.05)***	0.33(0.09)***	0.28(0.05)***	0.09(0.02)***	0.04(0.01)***	0.24(0.07)***	0.21(0.05)***
Journals and publications as important source of information for innovation	n.a.	n.a.	0.33(0.05)***	0.21(0.06)***	0.12(0.02)***	0.10(0.02)***	0.28(0.05)***	0.23(0.07)***
Government or public research institutes as important source of information for innovation	−0.02(0.008)*	−0.01(0.006)	0.14(0.07)	0.11(0.05)*	−0.06(0.02)*	−0.04(0.02)*	0.04(0.02)*	0.04(0.02)*
Universities as important source of information for innovation	−0.31(0.20)	−0.25(0.10)*	−0.15(0.06)**	−0.20(0.08)**	−0.26(0.12)**	−0.18(0.08)*	0.02(0.01)	0.01(0.004)*
Professional/industry associations as important source of information for innovation	n.a.	n.a.	0.21(0.09)**	0.29(0.13)**	0.15(0.04)***	0.19(0.07)**	0.37(0.16)**	0.42(0.18)**
Reliance on secrecy to protect innovation	0.04(0.03)	0.09(0.04)	n.a.	n.a.	n.a.	n.a.	n.a.	n.a.
Reliance on complex design to protect innovation	0.17(0.07)*	0.25(0.14)	n.a.	n.a.	n.a.	n.a.	n.a.	n.a.
Constant	0.65(0.41)	0.66(0.41)	0.45(0.27)	0.40(0.29)	0.40(0.31)	0.43(0.28)	0.50(0.31)	0.52(0.39)
N	64	64	138	138	181	181	213	213
Parms	15	12	29	12	46	12	65	12
RMSE	0.10	0.11	0.18	0.18	0.12	0.13	0.11	0.10
p	0.00	0.00	0.00	0.00	0.00	0.00	0.00	0.00
R-squared	0.16	0.15	0.20	0.18	0.15	0.15	0.11	0.11

* $p<0.1$; ** $p<0.05$; *** $p<0.01$

they will reduce their intramural R&D spending by a magnitude equivalent to 41 percent of their extramural R&D spending. On the other hand, if strategic coordination in Japan (0.77 on Hicks–Kenworthy) weakens to the level of its British counterpart (0.10 on coordination), Japanese firms will bring more innovation activities into their own hierarchy, increasing their intramural R&D by an amount equivalent to more than a third (36 percent) of their extramural R&D. Similarly, Italian firms (0.44 on Hicks–Kenworthy) focus more on conducting R&D in-house than their more coordinated (0.96) Austrian counterparts, devoting to intramural innovation an extra amount of spending equal to almost a third (29 percent) of their extramural spending. On the other hand, firms in strongly coordinated Finland (0.88 on Hicks–Kenworthy) carry out less in-house R&D than weakly coordinated Ireland (0.07): the ratio of intramural over extramural R&D in the former is smaller than the latter by 0.43, a difference equal to 34 percent of the average extent to which firms conduct R&D in-house rather than contracting it out.

This core pattern of finding is relatively consistent through the other surveys (2004, 2006, and 2008) reported in Table 5.1, which alternate between Hicks–Kenworthy and Hall–Gingerich as the measure of coordination. Looking across all four survey waves, coordination has a consistently negative effect on the ratio of intramural to extramural R&D. In other words, the more coordinated the economy, the less firms spend on internal R&D by subordinates relative to external R&D by independent contractors. Out of the eight SUR estimations in the table, only one fails to clear statistical significance for the coordination coefficient, and one clears on the 0.1 level. In most other estimations, the effect of coordination clears significance at least on the 0.05 level.

Depending on the year of the survey, one unit of increase in coordination on the Hicks–Kenworthy or Hall–Gingerich index will shave somewhere between 0.26 and 0.77 from the ratio of intramural over extramural innovation in the economy. A more realistically scaled one standard deviation change in coordination can still translate into a difference of up to 0.24 on the ratio, which is almost sufficient to transform an even split (i.e. ratio=1) between intramural and extramural innovation into a three-to-four mix in favor of the latter (i.e. ratio=0.75), or alternatively, make in-house R&D 25 percent more important than innovation by independent contractors (i.e. ratio=1.25), depending on whether coordination increases or decreases by one standard deviation.

Besides the quantitative evidence, qualitative evidence from the case study literature also provides concrete illustrations of how firms in liberal market economies rely on the formal hierarchy of ownership control, while firms in coordinated economies emphasize long-term relational subcontracting with other autonomous innovators. For example, Saka (2002) described the enormous difficulty faced by three British firms in their attempt to adopt the high-autonomy Japanese workplace management strategies during a period from 1998 to 2000. Two of these firms are British subsidiaries of Japanese companies (Teniki UK and Nissera UK, both of which are pseudo-names in order to protect the identity of the firms), and the other is an Anglo-Japanese collaboration (Rover-Honda). Although the actual outcome varied across these companies (worse in Teniki UK than the other two), all three more or less suffered the same problem: despite well-publicized "implementation," by the management above, of Japanese work systems, there was limited "internalization," by the workforce, of the high-discretion high-flexibility practice on the shop floor. Due to the lack of institutional resources for creating long-term relational commitment in the British economy, subsidiaries located in the UK were unable to replicate the positive synergy between high-trust, high-autonomy, and high-performance, which was typical of their headquarters in Japan. As predicted earlier in this chapter, without relational commitment, high discretion by the agent led not to high performance, but opportunistic exploitation. Saka provided a vivid portrait of how British workers took advantage of their new autonomy to engage in shirking: "Older workers at Teniki UK worked according to their own rules and enjoyed the freedom created by weak control in the factory. They manhandled machines when they did not work properly, ate and drank in their cells, and failed to fill in production timesheets on an hourly basis: 'I do it at the end of the day and take an advantage. It looks better that way' (assembler at Teniki UK)" (p. 260). Long steeped in a relationship regime marked by heavy management control, British employees were unflinching in their cynicism towards the sudden introduction of "the Japanese way": "We were forced to go on this course [on quality circles]. They called it 'family circle'. It is a big joke. Everything is a joke. It could be better if they were straighter with us. As far as we are concerned, they have deceived us... (operator in cluster assembly at Nissera UK)" (p. 260).

While firms in liberal market economies are marked by relatively rigid and hierarchical internal control, the case study literature highlights a very different form of relational governance for firms in coordinated economies: long-term cooperation among *autonomous* firms along the *vertical* production chain, in other words, *relational subcontracting*. For example, Tödtling and Kaufmann's (2002) survey of innovation practice in Upper Austria found a very strong emphasis by Austrian SMEs on relational collaboration among closely connected, local, subcontractors. When firms were asked to identify their main innovation partners, a total of 56 percent of them identified their "next-door neighbors" along the vertical production chain: customers or suppliers, which completely dwarfed the attention to other potential sources of innovation collaboration such as technology centers (5.5 percent), training institutions (8.2 percent), research organizations (5.5 percent), universities (6.8 percent), and public support (9.6 percent). Furthermore, such subcontracting ties are intimate and local: while 56 percent of Upper Austrian SMEs identified subcontracting partners from within Upper Austria, only 9.6 percent of them identified partners from the rest of the world. Such trust-based relational subcontracting was also evident in the region of Mirandola in Italy, which has emerged as a successful center of biomedical innovation. As Biggiero (2002) pointed out, in this dense social network of firms, long-term reputation is an important motivation that discourages the small subcontractors of large MNCs from engaging in short-term opportunistic behavior. In this particular context, reputation actually has two mutually reinforcing dimensions: first, small subcontractors invest in a reputation for reliability so as to sustain MNCs' willingness to contract them for the long term; second, long-term association with MNCs is especially valuable because of the positive "brand name" externality from MNCs' well-established international reputation. As noted earlier in Chapter 3, even when MNCs took over ownership of local firms in Mirandola, the foreign corporate headquarters left intact the entire local management structure and dense social network in Mirandola, imposing minimum hierarchical control. In other words, in their approach to relationship governance, these MNCs are Italian in all but name, and Biggiero (2002) aptly referred to such superficial ownership takeover as "pseudo-entry" by MNCs (p. 118).

Mirandola is part of a larger region in Italy's center-north-east, which consists primarily of SMEs engaged in flexible high-quality production. Known as the Third Italy, this region has a very different socioeconomic and industrial structure than Italy's traditional north-west industrial region (First Italy) and the low-skill labor-intensive southern region (Second Italy). The "three Italies" (Bagnasco 1977) underline the fact that the Italian model of capitalism can be understood as "regionalized capitalism," where the nature of socioeconomic coordination varies within country on a subnational level (Trigilia and Burroni 2009). A comparison between the First and Third Italy is especially interesting because they represent, respectively, the early and later engines of industrial innovation in contemporary Italy, and their comparison highlights the contrast between large corporate ownership and SME subcontracting as two different modes of relationship governance. The north-west region is where Italy's first industrial revolution began, and growing as it did during the early years of the twentieth century, firms in the region adopted the classic Fordist mode of production dominant at that time: vertical integration of multiple production divisions under common corporate ownership, formal and arm's length contracting relationship with other firms, and standardized production. Although the north-west region played an important role in the country's rapid economic growth in the 1950s and 1960s, the "conglomerate ownership" model proved ineffective in adapting to the transition towards high-quality, innovative, and flexible production. As a result, starting from the early 1970s, the north-west region was gradually eclipsed by the rise of Italy's center-north-east. Instead of large conglomerates, firms in the center-north-east are mostly SMEs, and instead of vertical *integration* through ownership takeover, production is carried out through vertical *subcontracting* between fully independent SMEs. Furthermore, instead of formal arm's length contracting typical of the north-west, subcontracting relationship in the center-north-east is deeply informal, social, and long term. Trigilia and Burroni (2009) highlighted various historical and organic features of the center-north-east region that helped generate a deep reservoir of social capital for relational subcontracting, such as dense networks of small and medium townships tracing back to medieval communes, and tightly knit regional "red" and "white" political subcultures associated respectively with Catholic and communist parties. The regional contrast between arm's length conglomerate ownership in the First Italy and relational subcontracting in the Third Italy is consistent with this chapter's overall theme that strong relational commitment brings high autonomy, while weak commitment leads to heavy ownership control.

Such subnational variation in capitalism exists not only within strategically coordinated European economies but also liberal market Anglo-Saxon economies. Marc Schneiberg (2007), for example, identified two very different industrial orders within the US economy. The first industrial order, similar to Italy's north-west, fits Chandler's (1977, 1990) classic description of the modern Fordist corporation, characterized by large vertically integrated corporate hierarchies and impersonal markets. The second industrial order, by contrast, contains various dense local institutions of nonmarket sociopolitical coordination. In other words, a "mini coordinated economy" and a "mini liberal market economy" coexist within the larger setting of the US as a liberal market economy. Consistent with this chapter's main argument, while the liberal market industrial order is characterized by classic Chandler-type conglomerate ownership control, the coordinated industrial order is a center of "alternatives to corporate hierarchy" (Schneiberg 2007, p. 57), based on *lateral* association of *autonomous* economic interests, such as electrical, dairy, and grain elevator cooperatives, property insurance mutuals, telephone mutuals, and state-chartered credit unions. In this "alternative industrial order," long-term relational commitment between autonomous firms along the vertical chain is much more common. For example, as Schneiberg pointed out, upstream dairy cooperatives and downstream chain stores such as A&P establish "insider status" relationship with each other, so that the upstream is guaranteed a long-term retail outlet while the downstream is guaranteed a steady source of supply. Geographically concentrated in a few states such as

Who Faces a Dilemma between Volatility and Output?

Nebraska, Kansas, Minnesota, Texas, and Iowa, this coordinated industrial order emerged with some unique conditions less relevant for European coordinated capitalism, such as the political influence of municipal public utilities and the Granger Movement. Nevertheless, its preference for relational subcontracting over vertical integration very much mirrors other coordinated subnational regions in Europe, such as the Third Italy and Upper Austria.

Overall, both the core quantitative evidence and the supplementary case study evidence are consistent with the chapter's main proposition that *control* (via the ownership hierarchy) and *discretion* (via independent contracting) are different strategies chosen by different types of capitalism, depending on the strength of strategic coordination. Both *control* and *discretion* may reduce the risks of opportunistic volatility in innovation. Nevertheless, just as firms wish to lower risks, they naturally also wish to raise output: besides low volatility in their revenue stream, firms also want high returns from their R&D efforts. As different strategies for managing opportunistic risks, do *control* and *discretion* affect innovation output in similar ways? As I show in the next main section of the chapter, although both *control* and *discretion* may smooth opportunistic risks, the former also carries the cost of lower innovation output because *control* crowds out the agent's motivation to raise output. In other words, the use of ownership control in liberal market economies to manage opportunistic volatility has a *demotivation effect* on innovation output. Firms in such economies face a binding tradeoff between volatility and output: the more they reduce volatility, the harder it is for them to raise innovation output. This *demotivation effect* vanishes in strategically coordinated capitalism, where principals grant *discretion* to agents as independent contractors. In such economies, there is no binding tradeoff: the goals of low volatility and high output are jointly achievable.

Who Faces a Tradeoff between Volatility and Output?

As a large theoretical and empirical literature has pointed out (Williamson 1985; Demsetz 1988; Falk and Kosfeld 2006; Bartling et al. 2012), although hierarchical "rule by fiat" may allow the owner to reduce the uncertainty of its agent's behavior, it also has the side effect of reducing the agent's output because *control* downgrades the agent's motivation. There are two channels though which *control* may demotivate the agent. First, it weakens the link between the agent's effort and reward, which reduces the power of incentives that the owner may offer to the agent (Milgrom and Holmström 1994; Holmström 1999). Second, it signals to the agent that the owner expects low agent motivation, which in turn discourages the agent from reciprocating with high efforts, completing a self-fulfilling prophesy of distrust (Luhmann

1968). In other words, while the former channel demotivates the agent by distorting its incentives, the latter demotivates the agent by distorting its beliefs.

Hierarchical *control* weakens the agent's incentives because, as scholars of contracts and property rights point out, the principal relies on the residual ownership of financial and physical assets used in production to set its boundary of control. Residual ownership gives the owner not only the exclusive right to decide how the assets should be deployed in circumstances unspecified in the contract, but also the claim to residual profits from these assets; that is, profits left over after compensation for other parties in the production (such as employee wages) are paid (Grossman and Hart 1986; Hart 1995; Barzel 1997; MacLeod 2002).

If the agent owns the financial and physical assets used in its own R&D work, it is an independent contractor rather than a subdivision of the principal. Residual ownership gives the agent full claim to the profits from creating the new technology and selling it to the principal. This ability to expropriate profits has two effects (Holmström 1999): on one hand, it gives the agent greater temptation to behave opportunistically outside the contract; on the other hand, it gives the agent greater incentives to work hard under the contract. When technological volatilities create conditions outside the contract's anticipation, they present opportunities for the agent to pass these technological risks onto the principal. Because the contract does not provide for how to evaluate the agent's action under these circumstances, the prospect of profit expropriation tempts the agent to act opportunistically so as to extract as much as possible from the principal. The previous section has discussed such agent opportunism in some detail: the agent, for example, can mislead the principal into agreeing to a later delivery that costs the agent less effort, or collect private rents from choosing undesirable suppliers, or exaggerate product quality so as to renegotiate higher prices. On the other hand, where the contract has made provisions for how to assess and reward the agent's work, the prospect of profit expropriation motivates the agent to work harder and raise output. In the words of Oliver Williamson (1985), the contract offers "high-powered incentives," in the form of residual profits. Therefore, while independent contracting may reduce the principal's capacity to *control volatility* under the ownership hierarchy, it may on the other hand help *increase output*, as a result of the high-powered incentives it offers to the agent.

By contrast, if the principal owns the financial and physical assets used in the agent's work (so that the latter is an employee or subdivision rather than an independent firm), the principal becomes the claimant to residual profits from the innovation project. Because now the agent no longer lays claim to the residual profits from its work, the incentives to work hard and raise output also weakens. Ownership *control* over the agent, in other words, lowers the power of incentives that the principal may offer to its agent. Of course, the principal can raise the power of incentives by handing over more residual ownership to its agent. However, as ownership transfers to the other side, the principal's residual *control* over the use of these assets also slips away, which weakens the principal's ability to suppress opportunistic volatility through "rule by fiat." In other words, hierarchical *control* creates a tradeoff between the principal's need to reduce opportunistic risks on one hand and the need to provide high-powered incentives for high agent output on the other. While *control* allows the principal to reduce opportunistic risks through "rule by fiat," it may also reduce output by demotivating the agent. This tradeoff inside the firm hierarchy nicely captures Milgrom and Holmström's (1994) notion of the firm as a "mixed blessing": it makes incentives less opportunistic, but also less powered (Holmström 1999, cited in Azoulay 2004).

Financial incentives such as residual profits are not the only motivation behind high agent output. As the literature on "prosocial" behavior has documented in numerous studies, individuals are often reciprocal in their behavior, responding to cooperation with cooperation, and to defection with defection (Levi 1988, 1997; Rolfe 2012; Frey 1997a; Fehr et al. 1998). In other words, a *trusting* gesture by one side may trigger *trustworthy* behavior on the other side, and conversely, the lack of "goodwill" by one side prompts the other to behave accordingly. By taking the agent's *discretion* away during innovation, hierarchical *control* sends a signal to the agent that the principal is *untrusting*, expecting the agent to have little motivation. The belief that the principal is untrusting, in turn, discourages the agent from reciprocating with high efforts, hence completing what Luhmann (1968) refers to as a "self-fulfilling prophecy of distrust" (Falk and Kosfeld 2006). This adds to a long-standing warning from the human resources management (HRM) literature that *control* carries its own cost (Lawler et al. 1995; Appelbaum et al. 2000; Ichniowski and Shaw 2003). Instead of distorting workers' explicit monetary incentives, the cost of control here is reflected in the distortion of workers' implicit beliefs about their principal's belief. As a result, Falk and Kosfeld (2006) refer to it as the "hidden" cost of control in their widely influential study. Falk and Kosfeld's experiments demonstrate that not only do agents reduce output in response to the principal's decision to exert control, agents also explicitly interpret the principal's use of control as a sign of the principal's distrust, which provides direct evidence for how the hidden cost of control arises. Recent experimental studies by Fehr et al. (2013) have documented further evidence that the lack of authority demotivates the agent emotionally, even if the agent is well aware that higher output is *better* for herself financially.

Of course, in reality, the explicit cost of control (in lowering the power of monetary incentives) and the hidden cost of control (in fueling mutual beliefs of distrust) do not work separately in demotivating the agent (Frey 1997b; Fehr and Rockenbach 2003; Ellingsen and Johannesson 2008). On the contrary, as Bartling et al. (2012) point out, the psychological and monetary effects of control in downgrading agent performance often work together. The authors' experimental studies identify a "dual world" of possibility for work organization, represented by two different environments: high and low trust. High-trust work environments are characterized by a combination of high agent discretion, substantial profit sharing by agents, and agent reputation. In return, agents deliver high output. Low-trust environments have the opposite combination: low agent discretion, little profit sharing by agents, and lack of agent reputation. In return, agents deliver low output.

These two regimes of high versus low trust map directly onto the contrast between strategically coordinated and liberal market economies discussed in this chapter. In liberal market economies, firms lack the institutional conditions for reputation and relationship building. Consequently, they cannot smooth opportunistic risks through long-term relational contracting. Without the prospect of long-term relational commitment, firms have to use *control* under the ownership hierarchy to prevent agent opportunism. However, the use of *control* lowers the power of monetary incentives and fosters mistrust, discouraging the agent from working hard to raise output. Because of this *demotivation effect* of *control*, firms in liberal market economies face a binding tradeoff between the need to lower volatility risks and the need to raise output from innovation.

Strategically coordinated capitalism, on the other hand, has the opposite combination. Because dense coordination institutions facilitate long-term relationship and reputation building, firms may reduce the risks of agent opportunism even if they leave *discretion* to their agents as independent firms, who become residual claimants to their own profit (Board 2011). These high-powered monetary incentives encourage the agent to work harder, as does the signal of trust conveyed by the principal's willingness to grant *discretion*, which encourages the agent to reciprocate with higher efforts. As a result, by combining long-term partnership building with the granting of full *discretion* to partners, firms in strategically coordinated capitalism do not have to face the dilemma between low opportunistic volatility and high output: these two objectives are jointly attainable. Since the volatility–output tradeoff is less binding in more coordinated capitalism, low volatility should interact with coordination in affecting innovation output: the more coordinated the economy, the smaller the negative impact on innovation output from reductions in volatility.

H2: *the more coordinated the economy, the smaller the negative impact on innovation output from reductions in volatility.*

Hypothesis H2 predicts that innovators in strategically coordinated economies face a less tight tradeoff between high innovation output and low volatility than their liberal market counterparts. However, there may also be an alternative explanation for this outcome: according to Peter Hall and David Soskice's (2001) classic argument, strategically coordinated economies focus on more incremental innovation, which is inherently more stable in its informational environment. Hence, the volatility–output tradeoff may be less binding for these countries simply because they specialize in "less volatile" technologies. This incremental-innovation-based alternative explanation, however, breaks down where it could have had the most bite: as will be shown later in the empirical analysis, the volatility–output tradeoff is weakest in some of the most strongly coordinated economies, in particular Nordic countries. Based on the alternative explanation, we should therefore expect these Nordic countries to engage in the least volatile, and most incremental, innovation. However, in reality, as Ornston (2012, 2013) and Kristensen and Lilja (2011) point out, it was precisely these Nordic countries that have successfully broken out of the traditional mold of incremental industrial innovation, and achieved breakthroughs in some of the most radical and volatile high-tech sectors, such as telecommunications. The fact that this Nordic success occurred in radical innovation (a sector where Anglo-Saxon economies are commonly understood to excel) is a strong indication that the better performance of coordinated capitalism in avoiding the volatility–output tradeoff is not confined to sectors of incremental innovation. In a similar vein, the case-study literature cited throughout the book also offers various concrete examples of European success in radical innovation, ranging from "middleware" software in Sweden (Casper and Whitley 2004) to biotechnology in Germany (Casper 2000) and Italy (Biggiero 2002). The example of German biotechnology may be especially revealing, because, as noted in Chapter 3, while the US continues to dominate the "therapeutics" segment of this sector, Germany has made a breakthrough in the "platform technologies" segment. As Casper (2000) pointed out, new knowledge created in the "therapeutics" sector tends to be formal and codifiable, making it easier for the principal (such as investors and management) to monitor the activities of agents (such as technicians and scientists). In the "platform technologies" sector, by contrast, knowledge is informal, tacit, and uncodified, making it much harder for the principal to monitor and discover agent opportunism. In other words, Germany's high innovation output occurred in a biotech segment marked by *greater* potential for opportunistic volatility than its US counterpart. This contrast further undermines the alternative explanation that coordinated economies avoided the volatility–output tradeoff simply by focusing on low-volatility incremental-innovation sectors.

Empirical Analysis

The key to hypothesis H2 is the interaction between volatility reduction and coordination. While coordination can be measured as before (alternatively through the Hicks–Kenworthy and Hall–Gingerich indicators), I have not yet discussed measures of volatility that are suitable for this chapter's context. As noted at the beginning of the chapter, the comparative political economy literature has examined economic risks as a form of "spread in variance" for many different economic outcomes that are pertinent to different stakeholders in society, such as workers (i.e. volatility in employment opportunities and wages), the government (i.e. volatility in macroeconomic outcomes and budget conditions), consumers (i.e. volatility in prices), or investors (i.e. volatility in stock markets) (Krause and Corder 2007; Scheve and Slaughter 2004; Hacker 2004; Doucouliagos and Ulubaşoğlu 2008; Béjar and Mukherjee 2011; Leblang and Mukherjee 2005; Kim 2007). Because this book's central theme is the sociopolitical coordination of businesses, I focus on the firm, and examine the volatility in private enterprise revenue. As profit-maximizing entities, firms not only have larger stakes in reducing the instability of their own revenue stream but also have greater power to do so, compared with instabilities in other broader macroeconomic conditions. In particular, as the focal question for this chapter, whether an individual firm uses *control* or *discretion* to deal with its opportunistic agent will have a direct effect on its own revenue, but unlikely to have a notable effect on the country's overall macroeconomic or fiscal conditions.

For data on private enterprise revenue, I turn to the OECD's Structural and Demographic Business Statistics data set (subsection Structural Statistics for Industry and Services). Based on nonpanel firm surveys, the data set provides an annual measure of private enterprise revenue, in millions of national currency, starting from 1998 for a large number of OECD countries. Unfortunately, similar to Eurostat's CIS, although the *information* about enterprise revenue is based on individual firm survey responses, the actual *data* are averaged to the sector level (up to ISIC Rev. 3 four-digit), providing a revenue estimate for an "average" enterprise in the given sector. As an estimate for the "typical" enterprise in the sector, this measure may have understated the full possible extent of volatility experienced by individual firms in the sector, because firms may be pushed to the tail of the revenue size distribution (and consequently weighted less in this central tendency measure) precisely as a result of having experienced high levels of volatility. Therefore, volatility calculated from this revenue measure should be regarded as a lower bound on the real possible extent of volatility experienced on the individual firm level. This will bias the empirical analysis against hypothesis H2: if there is evidence that reductions in volatility carry the cost of lower innovation output even when the real possible extent of change in volatility is underestimated, then the real tradeoff between volatility and innovation output can only be even stronger. Having explained the source data for enterprise revenue, now I discuss the estimation of volatility in enterprise revenue.

As is well known, GARCH (Generalized Autoregressive Conditional Heteroskedasticity) (Engle 1982) models may be used to model the effect of volatility directly from raw time series data. However, unlike other economic indicators commonly considered in volatility causal analysis (such as stock prices or consumer price indices), the enterprise revenue data have a much shorter time span of nine years (1998–2006) and much lower frequency (annual, as opposed to daily and monthly observations). As Lebo and Box-Steffensmeier (2008) point out, time series with very short spans or low frequencies are not appropriate for the GARCH family of causal models, such as DCC (Dynamic Conditional Correlations) used in analyzing monthly consumer sentiments or presidential approval ratings. Along the same vein, Leblang and Mukherjee (2004) suggest that valid inference from GARCH or EGARCH (exponential GARCH) models, or alternatively Markov-switching models, is confined to very high-frequency time series, such as (daily observed) stock market prices. Instead of directly *modeling the effect* of volatility from raw time series data using GARCH, I follow a simpler method that is less demanding on the limited enterprise revenue data at hand. First, I create *measures* of volatility from the data. Then, I use the volatility measure as an independent variable to estimate its effect, in particular, how it interacts with coordination in affecting innovation output (hypothesis H2). Without having to directly model the effect of volatility from raw time series data on enterprise revenue, I can avoid the use of GARCH methods that are not appropriate for the data's very short time span and low frequency.

As reviewed by Brunetti (1998) in detail, there are two distinctive methodological approaches to measuring volatility for time series economic data. The first approach is to calculate the standard deviation of the time series (Kormendi and Meguire 1985; King and Levine 1993). However, since the standard deviation is often strongly correlated with the arithmetic mean of the series, as a volatility indicator it will be contaminated by the size of the mean. In order to remove this bias, the standard deviation is normalized by its arithmetic mean, resulting in a widely used volatility indicator known as the Coefficient of Variation (Dollar 1992). In contrast to the Coefficient of Variation, Aizenman and Marion (1993) developed a different approach, exploiting the logic that volatility should imply a *deviation* of the series' *realized* future value from its future value *predicted on the basis of past value* (i.e. autoregressive prediction). On this basis, they first fit the time series with a first-order autoregressive estimation, and then use the standard deviation of residuals from the estimation as an indicator of volatility. For conciseness, I refer to this second volatility measure as Autoregressive Deviation.

To increase the robustness of findings, I adopt both approaches to measuring volatility. First, I use the nine years of revenue data to calculate a Coefficient of Variation as an indicator of revenue volatility over the nine-year (1998–2006) period. Second, following the Autoregressive Deviation approach, I fit the nine years of revenue data with first-order autoregression, and obtain the standard deviation of residuals as an alternative indicator of revenue volatility over this nine-year period. Since in both approaches the volatility indicator is calculated as a single statistic for the whole nine-year period, the indicator does not have cross-time variation. Instead, its variation is purely cross-sectional, across ISIC Rev. 3 sectors and countries. In other words, the variation in the volatility indicators captures how countries and sectors differ in the extent of revenue volatility experienced by their firms over the nine-year period between 1998 and 2006. As a result, the empirical testing of hypothesis H2 will also be framed in terms of cross-sectional variation: do low-volatility countries and sectors perform less well in innovation than their high-volatility counterparts? And is this volatility–output tradeoff more binding for less coordinated capitalism?

In this book, I have focused on *opportunistic* volatility; that is, volatility in firms' revenue stream caused by the opportunistic behavior of their agents. However, in reality, much of the volatility in enterprise revenue may not be opportunistic in origin at all, but driven instead by fluctuations in macroeconomic conditions. In order to parse out these nonopportunistic components of volatility from my volatility indicators, I regress my revenue volatility indicators on a set of variables capturing various types of macroeconomic volatility, and obtain the residual as the portion of revenue volatility cleansed off nonopportunistic macroeconomic fluctuations. More specifically, I regress the Coefficient of Variation for enterprise revenue on the volatilities of the following aggregate economic conditions, all calculated by the same Coefficient of Variation method: inflation, economic growth, the volume of stock market trade, unemployment, the size of private sector bank credit, the total number of patents granted by USPTO, and total R&D expenditure. After the regression, I obtain the residual volatility as the true measure of opportunistic volatility faced by firms. Similarly, I regress the other volatility indicator (Autoregressive Deviation) on the volatilities of the above said macroeconomic conditions, all calculated by the same Autoregressive Deviation method, and obtain the residual volatility. Compared with the raw revenue volatility indicators, the *residual* volatility indicators more precisely isolate the portion of revenue volatility firms face as a result of opportunistic behavior by their partners.

Hypothesis H2 suggests that while there may be a tradeoff between low volatility and high innovation output, how binding this tradeoff is will

Who Faces a Dilemma between Volatility and Output?

depend on how coordinated the economy is. The more coordinated the economy, the less binding this tradeoff, and hence the smaller the negative impact on innovation output from reductions in volatility. In other words, as the key outcome of interest from the interaction between volatility and coordination, innovation output will be the dependent variable for hypothesis H2. Starting from the third wave in 2001, each wave of CIS asked firms about the introduction of new products or production processes due to technological innovation. As I detailed in Chapter 3's analyses of product and process innovation, CIS measures product innovation through two indicators: the percentage of firms in a given sector reporting an increase in the range of products on offer, and the percentage reporting an increase in product quality due to innovation. Similarly, CIS provides two indicators for process innovation: the percentage of firms in a given sector reporting an increase in the flexibility of the production process due to innovation, and the percentage reporting a reduction in production cost due to innovation.

While both new products and new production processes are valuable output from innovation, Chapter 3's focus on jobs and productivity required that these two types of innovation output be clearly separated from each other, because they have different consequences (product innovation increases jobs while process innovation deepens productivity). In the current chapter, however, measuring one output at the exclusion of the other may no longer be appropriate. Since this chapter focuses on the tradeoff between volatility and innovation output, it is important to rule out the possibility that the finding is merely an artifact of countries differing in their innovation output by sector (i.e. comparative advantage in product versus process innovation). For example, if the data focus narrowly on process innovation, coordinated economies will automatically have larger innovation output, given their comparative advantage in such sectors. This large output would automatically make the volatility–output tradeoff look less severe in these countries, which biases the analysis in favor of the book's argument. On the other hand, when product innovation data are also included in the estimation, we are less likely to be biased towards finding large output (and hence small tradeoff) for coordinated economies. For the same reason, it may also be inappropriate to isolate the analysis to only data on radical (or incremental) innovation.

While the measure above focuses on the direct output from innovation (i.e. product and process improvements), CIS also provides another measure of innovation output, which pays more attention to its indirect manifestation in the firm's improved position in the market. Starting from the 2001 wave, each wave of CIS records the percentage of firms in a given sector that report having increased market share or entered new markets due to innovation. While "more market share" or "new markets" were asked together in the 2001, 2004, and 2006 wave, they were split into two separate questions in the

2008 wave. In order to maintain consistency with the earlier waves, I take the average response across the two questions for the 2008 wave.

In short, based on CIS data, I have created two sector-level measures of innovation output. The first measure focuses on what the innovation offers: the percentage of firms in a sector reporting the introduction of new products or production processes due to innovation. The second measure focuses on how the innovation is received by the market: the percentage of firms in a sector reporting increases in market share or entry into new markets due to innovation. Similar to other variables in CIS, the information on innovation output is reported on the sector level. The CIS sectors are classified according to NACE (Rev. 1/Rev. 2). Using the correspondence tables between NACE and ISIC classifications from the United Nations Statistics Division, these NACE sectors can be matched to ISCI Rev. 3 sectors, which are the units in the OECD data for the key independent variable (enterprise revenue volatility).

In total, four CIS waves (2001, 2004, 2006, and 2008) have sufficient data for the variables to be included in the analysis. Questions from the 2001 wave of CIS have a reference period of 1998–2000, and questions from the 2008 wave have a reference period of 2006–8. As a result, CIS recorded innovation output on four occasions during the nine-year period (1998–2006) over which enterprise revenue volatility is calculated: the 2001, 2004, 2006, and 2008 survey waves. I average the reported output across these four waves to obtain the mean innovation output over the volatility observation period. This mean innovation output over the period is the dependent variable for testing hypothesis H2. Besides direct information from firm surveys, innovation output may also be captured through the number of patents produced. The Eurostat's Science and Technology (Patent Statistics) database provides NACE/ISIC-level data on the number of patents granted by USPTO (per million inhabitants). As the dependent variable, I take the average of patents across 1998–2006 as the mean innovation output for this period, over which the revenue volatility statistic is calculated.

When innovation output is measured from CIS surveys (i.e. the percentage of firms in the survey reporting new products/processes from innovation, and the percentage of firms reporting increased market share/new markets due to innovation), I draw on CIS data to include several control variables that may have helped firms in the survey to raise their innovation output. Technical knowledge needed for innovation activities can be "embodied" in fixed investment (machinery or equipment) as well as "disembodied" in its raw form as pure information, passed on through R&D personnel, publications, or patent disclosures (Dollar and Wolff 1993; Pianta 1998; Spiezia and Vivarelli 2002). As a result, firms' total innovation output will be affected by how much they invest in embodied as well as disembodied knowledge. CIS provides, for each sector, enterprise expenditure (measured as percentage of the sector's total turnover) on the acquisition of machinery, equipment, and software. This will be a suitable control for investment in embodied technical knowledge. CIS also provides, for each sector, enterprise expenditure on the acquisition of external knowledge, which I adopt as a control for investment in disembodied technical knowledge. Of course, whether firms can effectively harness embodied or disembodied knowledge for their innovation task also depends on the expertise and skills of their R&D personnel. For this reason, I also include the CIS measure for the percentage of firms (of a given sector) engaged in the training of R&D personnel. Furthermore, firms may also carry out purely organizational changes that improve the production process without creating substantially new technological knowledge (such as the Just-In-Time practice of inventory stock management, and Quality Circle as a flexible environment for workers to communicate their individual suggestions for product improvement) (Kenney and Florida 1993). Although these innovative organizational practices are themselves non-technological, they may act as a catalyst for parallel development in technologies (Lazonick 2005; Piva et al. 2005). For this reason, I also control for the potential effect of organizational innovation on technological innovation output, using the CIS measure for the percentage of firms (of a given sector) that report having adopted new organizational strategies. Finally, I also control for total R&D expenditure by firms in the survey, measured by CIS as percentage of a given sector's total turnover.

Of course, when innovation output is measured as the number of patents from the Eurostat's Science and Technology (Patent Statistics) database, it is no longer appropriate to use those control variables from the CIS data, because the Eurostat's patent data is not based on the firms in CIS surveys. Correspondingly, I use various data sources outside CIS to control for the possible determinants of innovation output. To start with, I control for R&D expenditure (available from OECD's Science and Technology Database), fixed capital investment (from OECD Structural Analysis database subsection STAN Industry), stock market capitalization, the size of bank credit to private enterprises (from Beck et al. (2000)'s Database on Financial Development and Structure), as well as defense spending, all as percentage of GDP. Besides defense procurement, the state also invests directly in R&D, in order to provide various public goods and services for the economy. The knowledge spillover from government R&D may further help firms in their own innovation efforts (Nelson 2005; Freeman and Soete 1997). For this reason, I also control for R&D in the government sector. Furthermore, I also add a measure of government partisanship, so as to incorporate Roe (2002)'s finding that the redistributive demands of social democratic governments may weaken the entrepreneurial incentives of enterprises, and hence reduce innovation output. To measure government partisanship, I use the left party share of cabinet portfolio, available from Duane Swank's database Electoral, Legislative, and Government Strength of Political Parties by Ideological Group in Capitalist Democracies. With missing values for all variables considered, the analyses of patents cover thirteen OECD countries.[1]

Hypothesis H2 predicts an interaction effect on innovation output: the more coordinated the economy, the smaller the negative impact on innovation output from reductions in volatility. Since the interaction is between volatility *reduction* and coordination, I use the interaction term coordination* (–volatility), so that it directly captures how coordination moderates the effect of *reductions* in volatility. Since coordination cushions the negative impact on innovation output from reductions in volatility, the interaction effect on innovation output should be positively signed: the more coordination, the less damage to innovation output. I report the findings in Table 5.2, in which I measure innovation output using CIS firm survey data (percentage of firms reporting new products/processes due to innovation, or percentage of firms reporting increases in market share/new markets due to innovation).

As I noted earlier, both volatility indicators (Coefficient of Variation and Autoregressive Deviation) are calculated as a single statistic for the entire nine-year (1998–2006) period for which I have sector-level data on enterprise revenue. During this period, CIS recorded innovation output on four occasions (the 2001, 2004, 2006, and 2008 survey waves). I averaged the reported output across these four waves to obtain the mean innovation output over the volatility observation period. This mean innovation output over the period is the dependent variable. Therefore, although the analysis covers a multiyear period, the estimation relies on cross-sectional variation, identifying the interaction effect from the variation across ISIC sectors and countries in their revenue volatility and mean innovation output during this period.

In Table 5.2, odd-numbered columns use the Hicks–Kenworthy index to measure coordination, while even-numbered columns check the robustness of findings by switching to the Hall–Gingerich index. In Models 1 through 4, innovation output is measured as the percentage of firms in a given sector reporting the introduction of new products or new production processes as a result of innovation, while Models 5 through 8 measure innovation output through the percentage of firms in a given sector reporting increases in market share or entry into new markets due to innovation.

I start the discussion with Model 1, which measures revenue volatility as the Coefficient of Variation. Using this method, revenue volatility is calculated as the mean-normalized standard deviation in enterprise revenue. While the main focus is the interaction between volatility reduction (i.e. "–volatility")

---

[1] These countries are Australia, Austria, Denmark, Finland, France, Germany, Italy, Japan, the Netherlands, Norway, Sweden, the UK, and the US.

Table 5.2. The Effect of Volatility Reduction on Innovation Output (Community Innovation Surveys), 1998–2006

Volatility Measure	New Products/Production Processes from Innovation				More Market Share/New Markets from Innovation			
	Coefficient of Variation		Autoregressive Deviation		Coefficient of Variation		Autoregressive Deviation	
	(1)	(2)	(3)	(4)	(5)	(6)	(7)	(8)
Coordination*(−Volatility)	15.58(3.1)***	9.89(2.54)**	9.72(3.98)*	4.46(2.18)*	11.04(5.61)	13.82(2.06)***	6.21(1.39)***	9.11(2.27)***
−Volatility	−12.61(1.2)***	−8.59(1.2)***	−6.33(2.0)***	−2.29(0.63)***	−18.6(5.3)***	−11.8(3.4)***	−3.07(0.28)***	−2.75(0.8)***
Coordination	31.39(19.9)	8.24(3.81)**	21.04(13.6)	4.16(2.95)	24.61(15.07)	36.02(24.15)	18.7(2.82)***	6.16(4.33)
Expenditure on machinery, equipment, and software	4.69(1.08)***	6.51(1.30)***	3.02(0.79)***	3.98(1.15)***	8.36(1.81)***	4.79(1.21)***	2.04(0.59)***	2.51(0.71)***
Expenditure on acquisition of external knowledge	2.93(4.0)	2.15(1.84)	3.52(2.42)	1.62(1.24)	5.64(1.97)**	6.50(2.14)***	1.92(0.86)*	2.20(1.33)
% Firms training R&D personnel	3.04(0.99)***	3.59(1.15)**	2.45(0.56)***	1.11(0.20)***	2.79(0.93)***	1.04(0.29)***	4.01(0.97)***	3.15(0.82)***
% Firms adopting organizational innovation	1.18(0.48)**	0.56(0.31)	0.80(0.43)	0.63(0.40)	0.24(0.13)	0.47(0.25)	0.21(0.17)	0.14(0.08)
R&D expenditure (% of sector turnover)	5.76(1.03)***	3.14(0.87)***	3.98(1.35)**	4.27(1.26)***	4.62(1.31)***	4.08(0.97)***	3.97(1.0)***	5.80(1.79)***
Constant	16.06(14.8)	28.12(16.34)	19.22(10.0)	13.47(9.22)	23.19(16.77)	16.50(10.24)	30.43(14.51)*	12.59(8.43)
N	139	139	139	139	142	142	142	142
R-squared	0.14	0.12	0.19	0.15	0.14	0.16.	0.11	0.10

* $p<0.1$; ** $p<0.05$; *** $p<0.01$

and coordination, the "main effect" variable for –volatility is also of interest, because it captures the outcome when coordination is absent. This sets the stage for the interaction effect that will unfold with the onset of coordination. In Model 1, the "main effect" coefficient for –volatility is significant at the 0.01 level and negatively signed. In other words, when coordination is at its minimum (zero on the Hicks–Kenworthy index), each reduction in revenue volatility by one (mean-normalized) standard deviation will reduce by 12.61 the percentage of firms reporting new products/processes from technological innovation. This reduction in reported instances of innovation output is roughly equivalent to a quarter of the mean percentage (48.6) of firms reporting new products and processes across the countries. This suggests that, without institutions of sociopolitical coordination for businesses, firms face a substantively notable tradeoff between low volatility and high output: for each standard deviation reduction in their revenue fluctuation, firms pay the cost of a roughly 12 percent reduction in their chances for successful innovation output. However, this is very much the upper bound on the volatility–output tradeoff. As the statistically significant and positively signed interaction shows, the negative impact on innovation output from reductions in volatility is cushioned by the strength of coordination. The more coordinated the economy, the smaller the negative impact of volatility reduction. As Model 1 shows, a one-unit increase in the strength of coordination will reverse the negative impact on reported instances of innovation output by around 15.58 percent.

To illustrate the interaction, Figure 5.1 plots the marginal effect of –volatility on innovation output, conditional on the strength of coordination, with dashed lines making the 95 percent confidence interval. As the graph shows clearly, the negative impact on innovation output from reductions in volatility diminishes steadily as coordination gains strength.

When coordination is minimum at zero, each standard deviation reduction in the volatility of enterprise revenue comes with the cost of reducing the reported instances of new products/processes from innovation by 12.61 percent. This effect, of course, can also be directly recovered as the "main effect" coefficient for –volatility in Model 1 of Table 5.2. This negative impact on innovation output, however, starts to lose force as the economy becomes more coordinated. Almost half of this negative impact will be shaved off when coordination approaches 0.4 on the Hicks–Kenworthy index. At this level of coordination (close to Italy and France), the same increment of reduction in revenue volatility will only decrease the reported instances of successful innovation output by around 6.17 percent. When coordination further intensifies towards 0.6 on the Hicks–Kenworthy index (close to the levels of Belgium and the Netherlands), the negative impact of –volatility on innovation output drops entirely out of statistical significance. In other words,

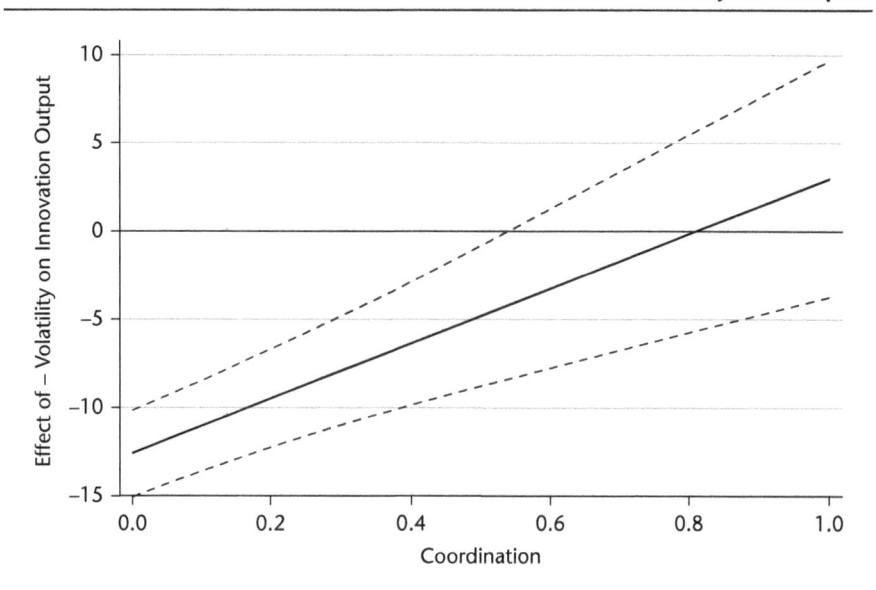

Who Faces a Dilemma between Volatility and Output?

Figure 5.1. Volatility vs. Innovation Output (CIS Surveys)

as long as the strength of coordination in the economy is somewhat above median (0.5 on the Hicks–Kenworthy index), the volatility–output tradeoff is no longer binding: better performance on one dimension (revenue stability) no longer results in worse performance on the other (innovation output).

To better appreciate the substantive implications of such interaction, I calculate the marginal effects (and associated standard errors) of –volatility on innovation output for different countries. For example, while each standard deviation decline in enterprise revenue volatility in weakly coordinated US (0.02 on the Hicks–Kenworthy index) will reduce the reported instances of new products/processes from innovation by 11.98 percent (standard error 1.23), the same increment of reduction in revenue volatility in better coordinated Italy (0.44 on Hicks–Kenworthy) has only about half the impact in suppressing innovation output, reducing the percentage of reported innovation success by 6.21 (standard error 1.77). Similarly, firms in weakly coordinated Ireland (0.07 on Hicks–Kenworthy) face a notably tighter volatility–output tradeoff than their counterparts in densely coordinated Germany (0.80 on Hicks–Kenworthy): while each standard deviation reduction in Irish enterprise revenue volatility comes with the cost of 12.42 percent fewer instances of reported innovation success, the same increment of revenue stability for German firms only reduces their reported instances of innovation success by 0.31 percent. With the standard error (2.7) ten times larger than the marginal effect, it is clear that in reality German firms do not face a statistically binding tradeoff between the need to lower volatility and the need to raise innovation output.

In other words, findings from Model 1 are consistent with the prediction from hypothesis H2: the more coordinated the economy, the less binding the tradeoff between low volatility and high innovation output. The tradeoff starts to drop out of statistical significance at a coordination level of 0.54. This implies that while Anglo-Saxon and Southern European countries (all of which have Hicks–Kenworthy scores below 0.5) have to *choose between* low opportunistic risks and high output, these two goals are *jointly attainable* for firms in Northern Europe and Japan.

Model 2 checks the robustness of this finding by using the Hall–Gingerich score to measure coordination, and again finds the same pattern: the "main effect" coefficient for –volatility is negatively signed and significant at the 0.01 level, indicating a very binding tradeoff between low volatility and high innovation output when coordination is absent; meanwhile, the coefficient for the interaction is positively signed and significant at the 0.01 level, indicating that this binding constraint slackens off as the negative impact on innovation output diminishes in more coordinated economies. As a different way to check robustness, Models 3 and 4 switch to another measure of enterprise revenue volatility: Autoregressive Deviation. To reiterate, the construction of this volatility indicator is based on the logic that volatility should imply a *deviation* of a time series' *realized* future value from its future value *predicted on the basis of past value* (i.e. autoregressive prediction) (Aizenman and Marion 1993). On this basis, I first fit the time series of enterprise revenue with a first-order autoregressive estimation, and then use the standard deviation of residuals from the estimation as an indicator of volatility. The findings from both Models 3 and 4 remain broadly consistent with earlier patterns: (1) the "main effect" coefficient for –volatility is negatively signed, indicating that, without sociopolitical coordination among businesses, the goals of lowering volatility and raising innovation output defeat each other; (2) the interaction is positively signed, indicating that as coordination becomes stronger, more of this negative effect on innovation output is reversed. For example, Model 3 shows that, without coordination, each standard deviation decrease in the Autoregressive Deviation of enterprise revenue will reduce instances of reported innovation output by 6.33 percent, but a one-unit increase in coordination more than compensates this negative effect, adding 9.72 percent back to the percentage of firms reporting innovation success. A similar pattern holds for Model 4, which measures coordination using Hall and Gingerich's data. Nevertheless, it is also important to note that when volatility is measured as Autoregressive Deviation (Models 3 and 4), the interaction effect clears statistical significance only at the 0.1 level; by contrast, when volatility is measured as the Coefficient of Variation (Models 1 and 2), the statistical significance of the interaction can reach the 0.01 level.

In Models 1 through 4, innovation output is measured as the percentage of firms reporting the introduction of new products or production processes due to innovation. This measure, as noted earlier, focuses on the immediate output from innovation, in terms of what the new technology "offers" (products and production processes). In Models 5 through 8, by contrast, I measure innovation output as the percentage of firms in a given sector reporting increases in market share or entry into new markets due to innovation. In other words, this second measure of innovation output places more emphasis on its indirect manifestation in the firm's improved position in the market, in terms of how the new technology is "received" by the market.

As columns 5 through 8 in Table 5.2 show, the core pattern of findings remains largely unchanged with this alternative measure of innovation output. Across all four models, the "main effect" coefficient for –volatility is consistently negative and significant at the 0.01 level, implying a binding volatility–output tradeoff in the absence of institutions that facilitate the sociopolitical coordination of businesses. On the other hand, the interaction between –volatility and coordination is consistently positive, which implies that coordination offsets the negative impact on innovation output. As a result, the volatility–output tradeoff becomes less binding in more coordinated capitalism. This interaction effect narrowly misses statistical significance in only one instance (Model 5). Overall, when coordination is measured using the Hicks–Kenworthy index, a one-unit increase in coordination may add 6.21 percent or 11.04 percent back to the reported instances of new markets or increased market share due to innovation, depending on the measure of volatility. Similarly, when coordination is measured using the Hall–Gingerich index, a one-unit increase in coordination will restore 13.82 or 9.11 to the percentage of firms reporting new markets or increased market share from innovation, depending on the measure of volatility.

Earlier in Figure 5.1 I graphed the interaction effect for innovation output as new products/processes and volatility as the Coefficient of Variation; now, in Figure 5.2, I use Model 7 to graph the interaction effect for a very different specification of these core variables: innovation output as increased market share/new markets from innovation, and volatility as Autoregressive Deviation.

The overall pattern in Figure 5.2 is very similar to the earlier Figure 5.1: the upward sloping line indicates that the impact of volatility reductions on innovation output becomes less negative as coordination grows in strength. When coordination is minimum at zero, each Autoregressive Deviation reduction in enterprise revenue volatility leads to the loss of 3.07 percent in the percentage of firms reporting innovation success. This negative effect, however, becomes smaller as coordination becomes stronger. By the time coordination reaches 0.32 on the Hicks–Kenworthy index, the confidence interval starts to straddle across zero, which indicates that the effect on

How Nations Innovate

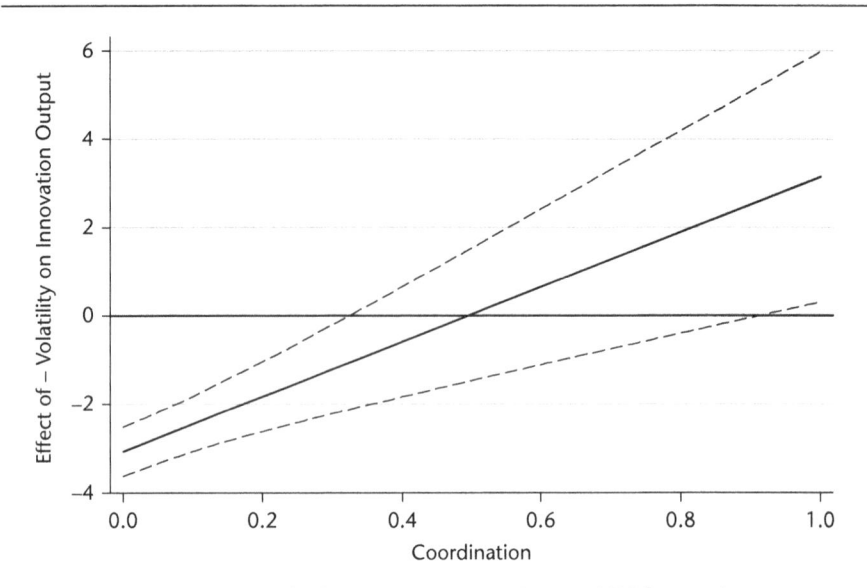

Figure 5.2. Volatility vs. Innovation Output (CIS Surveys)

innovation output is no longer statistically distinguishable from zero. The score of 0.32 on Hicks–Kenworthy neatly separates the Anglo-Saxon countries, all of which score below 0.20, from other European countries plus Japan, all of which score above 0.39. The fact that the confidence interval *crosses zero from below* at 0.32 on Hicks–Kenworthy is consistent with Hall and Soskice's (2001) suggestion that liberal and coordinated market economies, comprising respectively of Anglo-Saxon and European/Japanese countries, are two distinct worlds of capitalism with distinct logics for improving economic performance. While the Anglo-Saxon world is characterized by a *dilemma* between the need to lower volatility and raise innovation output, Europe and Japan are characterized by the *joint attainability* of both goals: raising performance on one dimension does not come at the expense of lowering performance on the other.

In fact, Figure 5.2 implies that the contrast between coordinated and liberal market economies may be even sharper than just suggested. At very high levels of coordination (above 0.92 on the Hicks–Kenworthy index), the confidence interval crosses entirely into *positive* territory. The three countries in this range (Austria, Sweden, and Norway) are well known to be some of the most strongly coordinated economies (Katzenstein 1985; Amable 2003; Pontusson 2005; Steinmo 2010; Ornston 2012). For these countries, the effect of reductions in volatility is to *increase* the percentage of firms reporting successful entry into new markets or increased market share due to innovation. The more firms increase the stability of their revenue stream, the greater their innovation output. As a result, what was a binding volatility–output

Who Faces a Dilemma between Volatility and Output?

tradeoff in liberal market Anglo-Saxon economies has become an active synergy in highly coordinated Nordic/Austrian capitalism.

While analyses to this point have measured innovation output using indicators from CIS, Table 5.3 reports findings for innovation output as the total number of patents granted by USPTO, measured on the NACE/ISIC-level. As the dependent variable, I take the average of patents across 1998–2006 as the mean innovation output for this period, over which the revenue volatility statistic is calculated.

In Model 1 of Table 5.3, volatility is measured as the Coefficient of Variation. For this model, the "main effect" for –volatility is negatively signed. Each standard deviation reduction in the Coefficient of Variation for enterprise revenue will reduce the number of patents by forty-eight per million inhabitants, which is almost a fifth (22 percent) of the mean sector-level patent output for countries included in the analysis. In other words, when innovation output is measured from patent data instead of firm survey data, there continues to be evidence for a substantively notable volatility–output tradeoff. The "main effect" of –volatility, of course, only captures the outcome when coordination is absent. As the positive interaction indicates, a one-unit increase on the Hicks–Kenworthy coordination index (equivalent to moving from minimum to maximum coordination) will be more than enough to completely offset the negative impact on innovation output, adding sixty-three back to the sector's number of new patents.

To illustrate this interaction, Figure 5.3 plots the marginal effect of –volatility on the number of patents, conditional on the strength of coordination. In the graph, the negative effect of –volatility on patents starts from the left at –48, and steadily diminishes in strength as coordination grows. When coordination reaches 0.41 (close to the level of France and Italy), the confidence interval crosses zero from below, and remains straddled across the zero line thereafter. In other words, when innovation output is measured by patents, the tradeoff between revenue stability and innovation output disappears after coordination reaches the (moderate) Southern European strength. Echoing the earlier finding from CIS (Figure 5.2), the finding on patents in Figure 5.3 again separates the countries into two distinct worlds of capitalism: a weak-coordination Anglo-Saxon world characterized by a binding tradeoff between low volatility and high output, and a strong-coordination European/Japanese world characterized by the joint attainability of these two goals.

Nevertheless, there is a subtle difference between the interaction slopes in Figure 5.2 and 5.3. In the former, the lower bound of the confidence interval climbs steadily with coordination, crossing zero and then entering positive territory. It is this persistent upward trend in the lower bound of the confidence interval that allowed the effect of –volatility to actually switch signs, flipping from a *tradeoff* (negative) under very weak coordination to a *synergy*

207

Table 5.3. The Effect of Volatility Reduction on Innovation Output (Patents), 1998–2006

	(1) Volatility as Coefficient of Variation	(2) Volatility as Autoregressive Deviation	(3) Italy/France dropped	(4) Australia Dropped	(5) Nordic dropped
Coordination* (-Volatility)	63.54(30.34)*	36.59(10.27)***	46.12(19.12)**	58.09(26.40)***	71.10(25.25)**
-Volatility	-48.06(15.01)***	-21.57(6.40)***	-37.14(11.15)***	-40.52(13.50)***	-31.23(7.48)***
Coordination	-3.34(4.35)	5.82(3.15)	5.62(4.77)	-6.10(5.27)	4.05(3.18)
R&D expenditure (% of GDP)	93.28(12.31)***	110.74(18.5)***	75.16(13.49)***	115.14(15.70)***	120.14(15.9)***
Fixed capital formation	6.26(2.84)	7.21(2.83)**	3.15(1.37)*	7.92(3.85)	6.24(1.91)***
Defense spending	41.82(12.53)***	25.23(6.57)***	32.61(11.04)***	39.20(12.36)***	25.84(7.50)***
Government R&D	120.86(27.95)***	97.46(18.60)***	130.52(36.51)***	125.07(34.16)***	95.14(20.57)***
Stock market capitalization	0.71(0.16)***	0.14(0.02)***	0.48(14)***	0.53(0.15)***	0.56(0.23)**
Bank credit	5.47(1.59)***	3.42(0.72)***	3.57(1.28)**	6.02(1.24)***	7.79(0.95)***
Left party share of cabinet portfolio	2.61(1.43)	1.04(0.49)	1.81(0.77)*	2.27(1.09)	0.98(0.55)
Constant	438.52(60.5)***	521.36(59.5)***	392.10(48.7)***	482.17(48.5)***	485.20(72.2)***
N	113	113	102	109	88
R-squared	0.74	0.69	0.67	0.66	0.70

* $p<0.1$; ** $p<0.05$; *** $p<0.01$

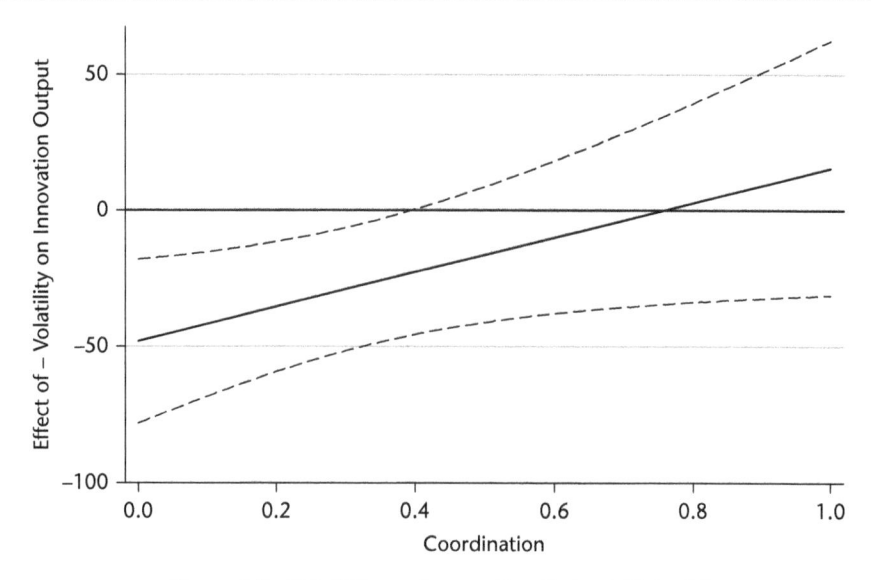

Figure 5.3. Volatility vs. Innovation Output (Patents)

(positive) under very strong coordination. In Figure 5.3, by contrast, although the upper bound of the confidence interval crosses zero from below early on (when coordination reaches 0.41), the lower bound never follows suit, experiencing some initial upward progression but then losing steam and falling back. In other words, in the case of patents, the force of interaction is never quite strong enough to transform the relationship between low volatility and high output from tradeoff to synergy. This relative weakness of the interaction effect is also reflected in the estimates from Model 1 of Table 5.3, where the interaction coefficient of 63.54 has a relatively large standard error of 30.34, clearing statistical significance only at the 0.1 level. Model 2 switches to Autoregressive Deviation as the measure of enterprise revenue volatility, with similar findings. While each Autoregressive Deviation reduction in volatility will take away twenty-one patents per million inhabitants in innovation output, moving from minimum to maximum coordination (one unit on the Hicks–Kenworthy indicator) more than reverses this effect, adding thirty-six back to the number of new patents produced by the given sector.

As I noted earlier in the book, the literature on varieties of capitalism has drawn increasing attention to the various "subtypes" of capitalism that exist within either liberal market or strategically coordinated economies. Some of these subtypes, such as Southern Europe and the Antipodes, are hybrid in nature, mixing symptoms of both strong and weak coordination (Amable 2003; Levy 1999; Culpepper 2003; Castles 1985; Watts 1997). The inclusion of such hybrid regimes may have suppressed the interaction between coordination and –volatility because, as Hall and Gingerich (2009) point out, hybrid types may be deprived of the institutional benefits from both strong and weak coordination. In other words, coordination in these countries may be both too weak (to sustain a high-trust *discretion*-based regime of risk management) and too strong (to trigger a low-trust *control*-based regime of risk management), in effect removing the role of coordination in how countries choose between *control* and *discretion*. In Models 3 and 4, therefore, I re-estimate Model 1 by respectively dropping France/Italy and Australia (New Zealand was never in the analysis, due to the lack of patent data), to see if the interaction effect is indeed notably stronger after these two potential "no effect" subcategories of capitalism are removed. As results from Table 5.3 show, the estimates for interactions in Models 3 and 4 are not notably different from Model 1. Along the same vein, I also consider the Nordic subset of coordinated capitalism, because these countries represent the very upper bound on the strength of strategic coordination, which may have disproportionately contributed to the interaction effect of coordination. In Model 5, I drop all Nordic countries from the analysis, to see if the interaction effect is indeed notably weakened after this set of very-strong-coordination countries is removed. As it turned out, the interaction effect in the "Nordic-excluding" Model 5 is in fact slightly larger than the "all-inclusive" Model 1. This indicates that the interaction effect found in Model 1 is clearly not due disproportionately to only a few countries on the upper extreme of coordination strength, but instead present across the broad range of coordination, as well as across different subcategories of capitalism.

In summary, this section has provided relatively consistent evidence that different types of capitalism face different worlds in managing the risks from opportunistic relationships: while firms in liberal market capitalism face a binding tradeoff between low opportunistic volatility and high innovation output, these two goals may be achieved jointly in strategically coordinated capitalism. Evidence for this contrast between liberal market and strategically coordinated capitalism was found for volatility measured either as mean-normalized standard deviation or as deviation from autoregression residuals, and for innovation output measured either from firm surveys (reported instances of new products/processes and new markets) or from patents.

Conclusion

Volatility is an important type of economic risk. When one firm supplies technologies to another, volatilities inherent to the innovation process present many opportunities for the agent to exploit the principal's disadvantage in information and extract additional surplus. As a result, *technical* volatilities faced by agents on the ground turn into *opportunistic* volatilities faced by principals. In order to reduce such opportunistic volatility, should the principal *control* its agent as a subordinate or grant its agent *discretion* as an independent contractor? How do varieties of capitalism differ in their choice between *control* and *discretion*? Between liberal market and strategically coordinated economies, whose innovators are "subordinates" under the ownership hierarchy, and whose innovators are "independent contractors"? These were the motivating questions for this chapter.

In this chapter, I have shown that firms in liberal market economies prefer *control*, and obtain technologies from divisions under their own ownership hierarchy; by contrast, firms in strategically coordinated economies prefer *discretion*, and obtain technologies from independent contractors. In strategically coordinated economies, dense institutions of sociopolitical coordination help firms establish long-term relational commitment. With relational contracting, the principal offers to renew the relationship repeatedly as long as the agent does not behave opportunistically. This prospect of long-term relationship discourages the agent from short-term opportunism even if the agent enjoys full *discretion* as an independent contractor (Chassang 2010; Lafontaine and Raynaud 2002; Klein et al. 1996). In other words, in economies with deep relational institutions, the principal can reduce opportunism even if it has no ownership *control* over its agent. Liberal market capitalism, by contrast, lacks institutions that facilitate long-term cooperation between firms. Without the prospect of long-term relational commitment, the principal has to use *control* under the ownership hierarchy to prevent agent opportunism (Lafontaine and Raynaud 2002; Klein 2002). When the principal chooses to have its R&D done by its own subdivision rather than an independent firm, it brings the agent under its ownership hierarchy. Residual ownership not only allows the principal to "rule by fiat" but also enables it to negotiate with greater bargaining power, both of which limit the agent's ability to opportunistically exploit its own informational advantage (Grossman and Hart 1986; Hart 1995; Barzel 1997; MacLeod 2002; Schmitz 2006; Lerner and Malmendier 2010).

Besides identifying *control* and *discretion* as different strategies to manage opportunistic relationships for different types of capitalism, I also compared how these two methods differ in their impact on innovation performance. Although *control* through ownership hierarchy may reduce opportunistic volatility, it also downgrades the agent's motivation to raise output. When the agent is not the residual owner of assets and profits from its own production, its incentives to raise output are weaker. *Control* through the ownership hierarchy, in other words, lowers the power of incentives that the principal can offer to its agent (Williamson 1985; Milgrom and Holmström 1994; Holmström 1999; Azoulay 2004). The principal may raise the power of incentives by handing over more residual ownership to the agent, but as ownership transfers to the other side, the principal's residual *control* also slips away, reducing its ability to "rule by fiat." As a result, hierarchical *control* creates a binding tradeoff between the need to lower volatility and the need to raise output: *control* allows the principal to reduce volatility through "rule by fiat," but it also reduces innovation output by demotivating the agent.

As a result, affluent capitalist economies become a "dual world" of possibility for relationship governance, represented by two distinct regimes: high and low trust. The low-trust regime is characterized by a combination of *control* and binding volatility–output *tradeoffs*; the high-trust regime is characterized by a combination of *discretion* and the *joint attainability* of low volatility and high output. While the first section of the chapter used CIS surveys to differentiate the two regimes on the *control/discretion* margin, the second section used a combination of survey and macro data on innovation output and revenue volatility to contrast the two regimes on the *tradeoff* margin. Based on various measures of volatility (mean-normalized standard deviation or deviation from autoregressive residuals) and innovation output (patents, firms' report of new products, new production process, and new markets/market share), I have shown that while reductions in volatility have a notable negative impact on innovation output in liberal market economies, this volatility–output tradeoff vanishes in strategically coordinated economies. In fact, in economies with very strong institutions of coordination, low volatility and high output may become *complements*: reductions in volatility have a statistically significant effect in *raising* innovation output.

6

Conclusion

This book has compared how varieties of capitalism differ in their patterns of technological innovation, and how this matters for various aspects of the economy and society. In technological innovation, new and incomplete information plays a fundamental role (Freeman and Soete 1997; Nelson 2005; Mokyr 2002). Therefore, by choosing innovation as the focus of comparison, this book was also a study in the political economy of information distribution. In this concluding chapter, instead of merely repeating the book's core messages, I reflect on some broader theoretical and practical issues that are logical extensions from the book's main arguments. First, I discuss the role of the state in the institutional transformation of coordinated capitalism, which complements the book's predominantly "firm-centered" theories of coordinated economies. In particular, I discuss how the state has affected the ability of coordinated capitalism institutions to adapt to a new, knowledge-intensive, economy. Then, I switch to a more practical perspective, and discuss various policy implications from the book's core findings, in areas ranging from financial market regulation to the design of welfare state and educational policies.

The Role of the State in Institutional Transformation

Throughout this book, institutions of coordinated capitalism have served as the master *independent* variable, driving the various differences in patterns of innovation across varieties of capitalism. However, in reality coordination institutions are themselves endogenous, slowly evolving in response to innovations in the economy. If we want to better understand the evolution of coordination institutions as a *dependent* variable representing conscious *institutional choice*, an exclusive focus on firms will no longer be sufficient, because the state also wields strong influence over institutional selection and design. The government's influence stems not only from its "final right," in the sense of the authority from being in power, to set public policies for the nation, but also from the unmatchable amount of resources it can marshal through government revenue and expenditure.

In this section, I provide a theoretical account of how the state has shaped the evolution of coordination institutions as coordinated capitalism adapts to the knowledge economy. As the discussion will show, the state not only helped firms reorient their innovation strategies, but also, in this process, gradually transformed the very function of some traditional coordination institutions, making them more compatible with the needs of a knowledge economy.

After the end of the Golden Age, coordinated economies faced growing challenges to their traditional mode of production based on incremental innovation and niche products. These challenges, however, turned out to be new opportunities for the state to assert itself in the economic landscape. In response, the state, especially in Nordic countries, assumed an important role on the supply side of the economy: the creation and preservation of high-quality human capital. The state implemented this new task of social investment through two channels. First, the state used *public funding* to support policy instruments that increase the *stocks* of high-quality human capital, such as early childhood education, daycare, active labor market training, sick pay, and formal education at all levels. Second, the state used additional *intervention* to reconfigure traditional coordination institutions so that they increase the *economic returns* of human capital stocks, in particular in the innovation of radical technologies and creation of knowledge-intensive jobs.

On the production front, coordinated economies had for many years consolidated an institutional foundation for comparative advantages in skill-intensive incremental innovation, in relatively mature and traditional manufacturing industries (Hall and Soskice 2001). However, the core manufacturing sectors went into decline and were eclipsed by the growing importance of radical innovation in knowledge-intensive sectors, such as telecommunications and biotechnology. From a traditional varieties of capitalism perspective, these countries appear ill prepared for this sectoral transition. Indeed, the focal institutions of coordination, such as patient capital and the internal labor market, all seem to impede radical innovation. However, as it turns out, in some countries the state played a crucial role in recasting the functions of coordination institutions, using existing institutional synergy to successfully break into radical innovation. These are precisely countries where political and popular support for the role of the state has been historically strong and enduring; that is, Nordic countries (Steinmo 2010). By contrast, coordinated economies with less state penetration (such as Belgium and the Netherlands) or with large public sectors but without Nordic levels of political support (such as Austria) failed to make a similar breakthrough as their mature industries declined.

Conclusion

It is no accident that the differing trajectories of Nordic versus continental countries are related to their differing welfare state models. The expansion of early childhood education and care in Nordic countries was motivated in the first instance by the goal of reconciling work and family and thereby increasing women's labor force participation. The universalization of general secondary school education and the expansion of tertiary education were motivated by the goal of increasing educational opportunity for the children of the less well off. In the course of the 1990s these quintessential social democratic policies, along with active labor market policies, were reconceptualized as social investments (Jenson 2011).

Kristensen and Lilja (2011) provide a detailed account of how Nordic countries transformed themselves from exporters of processed primary goods (agricultural goods for Denmark, wood products for the other three countries) to producers of high-tech goods and services, particularly in ICT. The ICT breakthrough was phenomenal in Finland and Sweden, resulting in annual per capita growth rates of over 3 percent in those two countries in 2000–7, well above the average 2.3 percent across all advanced economies. This success was not pumped by large increases in state spending on R&D. While there was a notable increase in aggregate R&D expenditure in Sweden, Finland, and Denmark, the state portion of this increase was five to ten times smaller (0.1 percent, 0.4 percent, and 0.2 percent of GDP respectively) (Stephens 2012). Norway, furthermore, did not experience more than marginal R&D growth in either the state or private sector, which according to Kristensen and Lilja (2011) is because continuing windfalls from oil exploration are delaying the moment when economic adjustment becomes imperative.

Without increasing public sector R&D funding, what did the state do to promote radical innovation? As noted earlier, while the Nordic state increased *stocks* of human capital through *funding* (more on this strategy later), it increased the *innovation returns* of human capital through policy *intervention* in traditional coordination institutions. Ornston (2012, 2013) characterizes this state-directed transformation of coordination institutions as "creative corporatism," in contrast to two other possible developments for coordinated economies, respectively "conservative corporatism" (which preserves preexisting coordination institutions geared towards incremental innovation in mature industries) and "competitive corporatism" (which dismantles coordination institutions to close the gap with liberal market economies).

How did Danish, Swedish, and Finnish firms and unions transform their coordination institutions from "conservative" to "creative" without drastically retrenching them in the direction of "competitive" corporatism? Although not implemented in reality, the threatening prospect of "competitive corporatism" as the ultimate consequence of reform inertia increased social partners' willingness to promptly embrace institutional adaptation. As

Ornston (2012, 2013) explains, this firm-level transformation was not possible without state-level political intervention by policy-makers. In all three cases, the state prodded economic actors to adapt their existing institutions to "creative corporatism," making a credible threat of otherwise unilaterally imposing cost-cutting reforms. Although the state encouraged both firms and unions to change, its intervention was particularly active with regard to organized labor. Just as the state used its own weakness in the international economy (for example, the growing constraint of the EU) as a credible commitment to reform, the growing weakness of unions (for example, Finnish unemployment was 17 percent in the early 1990s) prevented labor from credibly committing to resist reform. Instead, organized labor's rational best response was to accept investment-oriented new practices in exchange for some control over the content of these new practices.

Ornston (2012, 2013) argues that in moving to "creative corporatism," rather than abandoning coordination, the Nordic state took advantage of pre-existing stocks of cooperative apparatus to implement changes in three areas: finance, the labor market, and industrial policy. In finance, industry- and labor-managed pension funds were invested in venture capital markets to create institutional cooperation in early-stage finance of high-tech enterprises. In the labor market, skill formation took priority over job protection. More specifically, organized labor shifted from job security for core workers to collaborative skill investment for noncore workers. In industrial policy, the state took a conscious backseat, encouraging instead private sector interfirm cooperation in R&D. Such a turn towards high-tech competition would have been much more difficult without the existing institutional resources of coordination. For example, firms are willing to share private and sensitive information in radical R&D collaboration precisely because they have already had a long history of informational exchange in incremental R&D. Similarly, the state created research consortia, out of existing banking blocks and price-taking cartels typical of traditional coordinated capitalism.

One obvious question is why other coordinated economies failed to replicate the Nordic success. In essence, the set of conditions jointly necessary for "creative corporatism" is hard to find: (1) the state, being economically vulnerable, can credibly commit to reform (a condition absent in oil-rich Norway); (2) labor, being politically vulnerable to unilateral state intervention, cannot credibly commit to resist (absent in *Tarifautonomie* Germany or *Kollektivvertrag* Austria); (3) there is pre-existing strength in cooperation (absent in more conflict-ridden Belgium and the Netherlands); and (4) a pre-existing lead in social investment (absent in countries without social democratic welfare states).

Closely related to the transformation of production in coordinated economies is a transformation of employment. Kristensen and Lilja (2011), for example, attribute the Nordic success in innovation to fundamental changes in work organization, toward nonhierarchical, horizontally integrated, autonomous, flexible, and interactive practices. In other words, the Nordic pattern of work on the production floor has become increasingly characterized by "discretionary learning" (Arundel et al. 2007; Lundvall and Lorenz 2011), which involves high levels of problem solving and learning on the job and high levels of freedom for the worker to organize his or her work activity. In coordinated economies, what did the state do to increase the stocks of human capital that are suitable for high-quality knowledge-intensive employment? As I will discuss next, the state has increased public funding for a range of policies that boost human capital and employment in the knowledge economy. In other words, besides *policy intervention* to change patterns of *innovation*, the use of *public funding* to change the *quality of human capital* is another key element of the social investment state.

Nelson and Stephens (2012) examine the effect of public funding across seventeen OECD countries, directly on the quality of human capital (measured alternatively through years of education and performance scores in IALS), and indirectly through human capital on high-quality employment (measured alternatively as knowledge-intensive employment and discretionary-learning employment). The authors found that generous short-term unemployment replacement rates, high public spending on compulsory education, active labor market policy, daycare, and sick pay all serve to boost employment and, with the exception of sick pay, increase high-quality employment in particular.

These policies led to good employment outcomes through a causal channel based on the accumulation and preservation of skills compatible with knowledge-intensive employment. Daycare, as a key instrument in early childhood education and care (ECEC), not only helps mothers avoid skill atrophy by helping them remain in the labor market, but also helps children improve those cognitive skills that are now recognized in the literature as crucial for employment in the knowledge-intensive sector (Broberg et al. 1997; Campbell et al. 2001; Waldfogel 2002). Beyond ECEC, the content of compulsory education has increasingly moved beyond basic writing and numeracy to areas crucial for the knowledge-intensive service economy, such as creative development, life-long learning, technological and scientific knowledge, and employability (Benavot and Braslavsky 2007). While daycare and education spending accumulates skills primarily for the young, other policies protect and deepen skills for working adults. For example, through retraining, active labor market policies not only prevent skill atrophy but also facilitate further skill acquisition for the unemployed. High but short-term unemployment benefits prevent the unemployed from being forced, by financial pressure, to take on jobs that do not preserve or deepen their human capital. Similarly, sick pay prevents workers from being forced to return to work before complete recovery, so that suboptimal utilization of human capital on the job can be avoided (Huo et al. 2008).

Public expenditure on these social investment policies has important implications for the long-term job creation prospect of coordinated economies. As the literature has long pointed out, with the transition to the service economy, relatively strong wage compression in coordinated economies may prevent them from adapting effectively to the decline of manufacturing (Iversen and Wren 1998). However, some coordinated economies outgrew these dim expectations. Where political and public support for state funding is strong (again, Nordic countries), the state used the above-mentioned social investment policies to increase service sector employment, not at the low end, but at the knowledge-intensive high end characterized by discretionary learning. As Huo and Stephens (forthcoming) discuss in detail, the Nordic countries lead in the percentage of workers in discretionary-learning jobs, followed by continental European countries, with Anglo-Saxon economies trailing behind. This is all the more striking given that Anglo-Saxon countries actually do very well in overall employment when the quality of employment is not taken into consideration.

In summary, discussions above suggest that, when we conceptualize coordination institutions as endogenous and subject to conscious design and selection, the active role of the state becomes indispensible in understanding the transformation and innovation of coordinated capitalism. As this book draws to a close, I use its final pages to discuss some practical public policy implications. I organize the policy topics into the following three themes: (1) financial market regulation, (2) workforce productivity, and (3) social policy.

Financial Market Regulation

Should Financial Markets Be More Transparent?

At first sight, this may appear to be a superfluous question. After all, who can justifiably ask for more opacity on the financial market? More public disclosure of information, one may reasonably argue, will allow more investors to assess the quality of projects more accurately, and hence reduce the chance of low-quality projects being funded. By diminishing the future risk of another widespread financial failure, public disclosure presumably can only work to the benefit of investors and financial industries alike. In this vein, stricter requirements on public disclosure have become a key reform agenda for the G20 in the wake of the recent financial crisis. For example, the organization's Washington Declaration in 2008 called for greater financial market transparency (especially for complex products), and in a related move, the Financial

Stability Board and the International Monetary Fund (2009) highlighted the role of "information gaps" in the financial crisis. Similarly, the EU also responded to the crisis by introducing new public disclosure rules on capital levels (Capital Requirements Directive), credit rating agencies (Credit Rating Agencies Regulations I–III), net short positions (Short Selling Regulation), and alternative investment fund leverage (Alternative Investment Fund Managers Directive), among others (Moloney 2012, pp. 170–1).

Nevertheless, despite the apparent surge in popularity for measures to enhance financial market transparency, more public disclosure of corporate information may not necessarily work to the advantage of all investors. A key implication from Chapter 2 is that different types of information may crowd each other out: since the same hidden knowledge cannot be revealed both indiscriminately and selectively, there is a tradeoff between public and insider information. The more hidden knowledge is released in one form, the less is available in the other. As a result, while *public disclosure* of private knowledge is commonly understood to enhance (public) information on the financial market, the *selective withholding* of knowledge, somewhat counterintuitively, may enhance information of another type: insider information. Therefore, depending on whether investors rely on public information (such as stock market investors) or insider information (such as banks and private equity investors), they may benefit either from public disclosure, or a *waiver* on public disclosure.

For this reason, instead of "uniform," the enforcement of market disclosure should be "targeted," seeking different treatments for markets that consume different types of information. As Chapter 2 pointed out, while the communication of insider information signals relational commitment and serves as an important source of relational *stability*, the indiscriminate release of public information is especially effective in pooling *risks* across the market. Policy-wise, this implies that strong (weak) enforcement of public disclosure may be more appropriate for markets that put greater priority on risk-seeking (stability). This logic is consistent with the overall principle of financial regulation spelled out by Ethiopis Tafara, the former Director of International Affairs of the Security Exchange Commission (SEC). As Tafara (2012) pointed out, while the regulation of banking and insurance (which prioritize stability) should be discrete and behind-the-scene, the regulation of securities (which prioritize risk-taking) should be aggressive and in the full public view. This differentiated approach to financial regulation may help preserve both insider and public information where they matter most. Because banking and insurance markets rely heavily on insider information, carrying out regulation behind the scene will help prevent the insider information revealed during regulation from being dispersed to the public (and hence losing all its value). Conversely, because the securities market relies heavily on public information, enforcement in the full public view will allow hidden information to be revealed broadly to all investors (and hence maximize its value).

Against the backdrop of this broad regulatory principle, we can also identify specific policy examples that appear to be inconsistent with the informational nature of the market at which the policy is targeted. For instance, Articles 5 and 6 of the EU's Alternative Investment Fund Managers Directive set a lower notification threshold for *reporting to competent authorities* than for *public disclosure* of net short positions in equities. This, in effect, creates insider information (observable to competent authorities but not the public) on an equity market that otherwise relies heavily on public information. These pockets of insider information, as a result, may prevent the equity market from effectively pooling risks across the millions of individual investors. Similarly, under the EU's Transparency Directive Proposal, investors with large shareholdings face tougher requirements than regular investors to disclose trading positions to *both regulators and the market* (in other words, making their trading positions public information) (Moloney 2012, pp. 171–3). Since investors with larger stakes are more likely to be patient "insider" capitalists who prefer "voice" to "exit," these measures may undermine the very informational basis for patient financing relationship, by forcing patient capitalists to divulge their insider information to the public.

Besides the type of *financial markets*, public disclosure policies can also be evaluated for their consistency with the *type of countries* in which they are pursued. For example, an important objective of Rule 144A of the SEC was to promote venture capital funding from institutional investors. Under Rule 144A, these institutional "venture investors" are exempted from the requirement to make certain public disclosures, hence allowing them to preserve insider information (Tafara 2012, pp. xviii–xix). Since, according to Chapter 2, liberal market economies tend to lack insider information, and as the chapter demonstrated, their venture capital performance suffered as a result, the effect of Rule 144A in protecting insider information appears to be consistent with its objective of fostering more effective venture capital in the US.

How to Foster Financial Markets in Europe?

A key finding from Chapter 2 is that, because different types of capitalism specialize in communicating different types of information, they also excel in different types of financial markets. By encouraging the dissemination of public information, liberal market economies do well in the stock market; by encouraging the preservation of insider information, strategically coordinated European economies do well in banking, and somewhat surprisingly, venture capital finance. As examples of respectively "debt" and "equity," banking and

Conclusion

venture capital cannot be more different from each other as financial instruments. Nevertheless, European economies enjoy high performance in both these two markets, because both rely heavily on insider information. Can European governments build on this insider information advantage, and further stimulate the growth of both markets?

As I noted in Chapter 2, compared with bank lending, venture capital is a much less self-contained source of finance. Instead of being terminal, venture capital is only the very first stage of a multistep funding process, after which the vested entrepreneur must be able to continue its flow of funding through other means (such as buyouts and IPOs) in order to remain viable. While buyouts are similar to venture capital as examples of private equity provided by insider-knowledge investors, "going public" through IPOs relies on the flow of an altogether different type of information across the stock market: public information.

As Black and Gilson (1998) pointed out, a well-developed stock market that allows venture capitalists to exit profitably through IPOs is crucial for the very development of the venture capital market itself. When the venture-backed entrepreneur goes public, the dilution of ownership control among millions of individual investors on the stock market *increases the entrepreneur's* own control rights *relative to investors*, hence allowing the entrepreneur to *regain control* over the firm, which was initially ceded to venture capitalists in order to secure their funding. This benefit cannot be realized through other theoretically possible exit mechanisms such as buyouts and bank finance, because both simply transfer majority control rights from one set of alien principals (venture capitalists) to another (banks and buyout capitalists). Furthermore, the option of "going public" is only available to the entrepreneur if she has indeed delivered high performance. As a result, compared with bank or buyout finance, IPOs are far more effective in enforcing an "implicit contract" between entrepreneurs and venture capitalists: the former promise to deliver high performance, in return for which the latter promise to relinquish control. Consistent with this theoretical argument, the empirical literature has found strong evidence that venture capital exits through IPOs are more profitable than through other means. Gompers (1995), for example, found that IPOs-based exits generate around 60 percent annual returns on venture capital investment, while other exits provide only around 15 percent. Similarly, Black and Gilson (1998)'s comparative study found that, in countries where stock markets are large (the US, the UK, and Israel), venture capital markets are also large, while in countries where banks are dominant (Germany and Japan), venture capital investment is trivial.

These insights, combined with the book's findings on insider and public information, have some implications for how European governments may more effectively foster venture capital and banking. Although the two markets both benefit from the rich insider information in Europe, they may not lend much real synergy to each other, because bank-backed finance cannot serve as an effective exit mechanism for venture capital finance. The real complement to venture capital lies instead in IPOs, on the stock market. However, IPOs and venture capitalists thrive in opposite informational environments: the former rely on public information while the latter rely on insider information. In other words, to effectively stimulate the further growth of European venture capital, European governments must look to an area where their economies have traditionally suffered a (public) information disadvantage: the stock market. To overcome the relative paucity of public information without compromising the depth of insider information in Europe, government regulation of information disclosure by *firms* needs to be as subtle and targeted as the policies for disclosure by *financial services* outlined in the previous section by Tafara (2012). In other words, instead of a uniform approach that requires all firms to disclose to investors under all circumstances, information disclosure should be conditional. Conditionality provides policy-makers with the discretion to accurately discriminate between those contexts where public information is important and where insider information is important.

For example, because the required informational environment shifts from insider to public information when high-tech startup firms make the transition from venture to IPO financing, one option to "contextualize" disclosure is to condition disclosure on *financing stage*. Since public disclosure is less important before firms issue IPOs, and since once disclosed information cannot be "unlearnt" by investors, European governments may adopt a disclosure regime that tightens by stage, so that firms face strong disclosure requirements only *after* they successfully complete the venture capital stage. Under this conditional disclosure regime, a firm's information is only made public *after* its "insider value" has been properly appropriated by "insider investors," so that the revelation of public information and preservation of insider information do not crowd each other out. Of course, such conditional enforcement of public disclosure may also help liberal market economies improve their area of weakness (venture capital) without compromising their area of strength (stock markets). More generally, these possibilities to lessen the tradeoff between insider and public information highlight that the economic effect of capitalism institutions should not be understood as a deterministic force, even though the book has gone to great lengths in emphasizing the role of coordination (or its absence) in molding economic activities. The possibility to overcome institutional constraints through flexible policy adjustments is a theme that will appear again in my later policy discussions.

Since established firms in mature technologies tend to rely less on the venture–IPO nexus for financing than high-tech startup firms, another option to "contextualize" public information disclosure is to condition disclosure *on sector*. As noted earlier in the book, the OECD (Hatzichronoglou 1997) has established a detailed classification of industrial sectors by technological intensity (high, medium-high, medium-low, and low), using NACE (Rev.1) and ISIC (Rev.2). Using this highly disaggregated (up to three-digit NACE) scheme, European governments can effectively identify individual firms' technological intensity, and correspondingly impose less stringent disclosure requirements on less technologically intense firms. Since such firms tend to rely more heavily on banks for external financing, less aggressive public disclosure may help these firms more effectively establish deep "insider" relationship with their investors, shielding them from the more indiscriminate revelation of information demanded of higher-tech firms raising money on the stock market.

Workforce Productivity

How to Make Manufacturing Jobs More Productive?

The comparative political economy literature has long recognized that Anglo-Saxon workers lag behind their European/Japanese counterparts in productivity on the shop floor. So far, the explanation for this productivity deficit has focused on the lack of deep skills and ineffective organization of the workplace. In return, the solution to the Anglo-Saxon productivity problem has focused on strategies to encourage asset-specific skill training, as well as more autonomous workplace organization (Kenney and Florida 1993; Soskice 1994; Crouch et al. 1999; Culpepper 2003; Thelen 2004; Martin and Swank 2012).

However, deeper training of skills or better organization of work may not be sufficient. The marginal contribution of human capital to productivity diminishes over time, unless this decline is offset by simultaneous innovation in technologies (Solow 1957; Jorgenson 1995; Rosenberg 1996; Lau 1996; Griliches 2000). Not all types of innovation are equally effective in boosting productivity. As Chapter 3 has shown, compared with process innovation, product innovation has much weaker impact in raising productivity on the shop floor. As a result, by specializing in product innovation, Anglo-Saxon economies appear to have the "wrong kind" of innovation for the purpose of productivity improvement. What public policies may encourage Anglo-Saxon firms to engage in more process innovation?

As noted in Chapter 3, product and process innovation lead to very different types of competition on the product market. Product innovators compete through *differentiation* on the product space, which reduces product substitutability and dampens price competition. As a result, the market may be segmented into multiple firms each having a positive market share. Process innovators, by contrast, engage in the opposite type of competition: instead of differentiation in products, they compete in production efficiency (and hence cost and pricing). With more substitutable products and fiercer price competition, a technologically efficient low-cost entrant can displace an inefficient high-cost incumbent, creating the possibility for market monopoly (Raith 2003). The very prospect of this monopoly based on winning the efficiency race may, therefore, spur firms towards process innovation.

This implies that, for Anglo-Saxon governments, a nuanced approach to "predatory pricing" may help them in their battle to improve workforce productivity. In particular, as Besanko et al. (2014) highlighted, sometimes lower prices do not just prompt the exit of rivals from the market; they may also be a manifestation of the incumbent's success in "moving down the learning curve," that is, successful innovation in process technologies that help improve cost efficiency. Therefore, Anglo-Saxon governments will benefit from a measured (as opposed to doctrinaire) response to predatory pricing, distinguishing, for example, whether lower prices represent a "race to the bottom" (lower wages, poorer working conditions, and lower regulatory standards) or a successful innovation in process technologies. Do Anglo-Saxon countries present the right political conditions for this type of sympathetic view towards price competition?

As Chang et al.'s (2011) recent influential work pointed out, majoritarian electoral politics in Anglo-Saxon countries is indeed more favorable to price competition than proportional-representation Europe. According to the authors, consumers are normally disadvantaged relative to producers in politics, because while the former have only votes, the latter can use both votes and financial contribution to influence government policies. However, the first-past-the-post electoral system results in seats that are more marginal, whereby the marginal utility of a single voter is larger (from the perspective of the politician). As a result, Anglo-Saxon politicians will be under greater pressure to favor consumers at the expense of producers than their European counterparts. Consistent with this logic, the authors showed that product prices are systematically lower in Anglo-Saxon economies than Europe. In other words, just as the goal of higher workforce productivity in Anglo-Saxon countries may call for greater government tolerance towards price competition, electoral politics in these countries also predisposes their governments towards this policy stance.

How to Create More Enthusiastic Workers?

High performance at work depends not only on worker skills. A highly skilled agent will not produce high output if she is not *motivated* to put in the effort. In other words, beyond the agent's *expertise*, effective employment policies should also strengthen the agent's *incentives*, that is, create more "enthusiastic" workers. What kind of work environment is more conducive to highly motivated workers? This is one of the most important topics not only in popular books but also the HRM literature. As many scholars highlighted, too much *control* demotivates the workforce (Lawler et al. 1995; Appelbaum et al. 2000; Ichniowski and Shaw 2003). Too many rules and constraints over workers' behavior may lead them to infer that their employer is untrusting, which in turn prompts reciprocal untrustworthy behavior from workers (Falk and Kosfeld 2006). High-performance work environments, instead, should give workers greater *discretion*. As the literature on discretionary-learning employment points out, the Nordic success in creating a highly motivated and productive workforce in knowledge-intensive jobs is strongly related to fundamental changes in work organization, toward non-hierarchical, horizontally integrated, autonomous, and interactive practices (Arundel et al. 2007; Lundvall and Lorenz 2011). In these high-motivation work environments, workers' tasks involve high levels of problem solving and learning on the job, and they enjoy high levels of freedom to organize their own activity.

As Chapter 5 has shown, firms in liberal market Anglo-Saxon economies tend to rely more heavily on *control* to suppress agent opportunism in potentially volatile relationships (such as during technological innovation). *Control*, in turn, crowds out agents' motivation to raise performance. Firms in coordinated European economies, by contrast, grant agents more *discretion*, which in turn motivates higher agent performance. As a result, the two varieties of capitalism represent two distinct worlds of workplace organization: a high-control/low-performance Anglo-Saxon regime, and a low-control/high-performance European (and especially Nordic) regime. Given this contrast, can the US (and other liberal market economies) narrow the gap, by granting more autonomy in the workplace?

Greater workplace autonomy alone may not be sufficient for the US economy to replicate the Nordic success. According to Chapter 5, what underpins the high-motivation work environment in Nordic economies is a *combination* of high autonomy and long-term relational commitment. As Bartling et al. (2012) pointed out, high agent discretion delivers high performance only when the agent cares strongly about the long-run continuation of the relationship. In order to attract "repeat customers," the agent invests in a reputation for being reliable, and refrains from exploiting her own discretion for short-term gains. When there is no relational commitment, by contrast, agent autonomy makes things *worse* for employers. As Bartling et al. found, when agents do not care about establishing reputation, they take advantage of their own autonomy and shirk even more, driving down output. For this reason, to characterize the Nordic high-performance work environment as "high autonomy" may have missed the forest for the tree. At the heart of the difference between the European and Anglo-Saxon workplace settings is relational commitment: one is a high-trust regime, while the other is a low-trust regime.

As noted in Chapter 5, although the US economy as a whole is commonly understood as an example of liberal market capitalism characterized by low relational commitment, Marc Schneiberg (2007) has highlighted considerable subnational variation in the nature of relationship governance. In particular, a "mini coordinated economy," geographically concentrated in a few states such as Nebraska, Kansas, Minnesota, Texas, and Iowa, can be identified as an alternative industrial order, based on long-term relationship commitment between smaller firms, autonomously organized through lateral associations such as credit unions, mutuals, and cooperatives. Furthermore, Chandler-type conglomerate ownership control is also less common in the internal structure of these firms, making them a source of "alternatives to corporate hierarchy" (Schneiberg 2007, p. 57). Combining high relational commitment with low control, this alternative industrial order may be better positioned to foster a motivated workforce than is commonly expected from a purely national-level understanding of the US economy.

Relational commitment is not the only possible mechanism for motivating high agent efforts. An important alternative to the long-term *building of relationships* is the *powering of incentives* (Lafontaine and Raynaud 2002). Even if the agent does not care strongly about the future renewal of the relationship, as long as she is given a sufficiently high stake in the principal's output (for example, owning a share of the principal's profit), she will have sufficiently "high-powered" incentives (Williamson 1985) to refrain from opportunistic actions that may reduce output. Examples of such Employee Stock Ownership Plan (ESOP) are not hard to find in the US (such as W. W. Norton & Company, Schreiber Foods, Hensel Phelps Construction, and Avis Rent a Car System (until 1996)), and ESOP companies are not shy to advertise their worker-owned nature (for example, on Hy-Vee's very own logo). As Dow (2003) pointed out, the real motivating effect of worker share-ownership depends on the size of the work team. When profits are dispersed across a very large number of workers, an additional collective action problem appears: even if each worker has a high *stake* in the firm's total output, her own *influence* over this output is too small to be pivotal. For this reason, high-powered incentives may motivate employees more effectively in "legal partnerships" characterized by small-team profit-sharing, such as law, consulting, and accounting firms (Kaya and Vereschagina 2014). Because such small-team firms tend to rely disproportionately more on advanced human capital (such as academic education and client knowledge) than machinery (Rajan and Zingales 2001), Chapter 4's findings on overeducation in Anglo-Saxon economies imply that these economies provide a good environment for such employee-motivating firms to thrive.

Conclusion

The preceding discussion suggests that, in order for the US to emulate some of the Nordic success in creating highly motivated employees, a virtual "regime change" (which fundamentally reconfigures American liberal market institutions towards coordinated capitalism) may not be necessary. In other words, the effect of national-level coordination institutions (or their absence) on economic performance is not deterministic, but open to flexible adjustment, a theme already noted earlier in this chapter's discussion of financial market regulation. In the US, there are already existing regions (as in Schneiberg's alternative industrial order) or firm structures (such as legal partnerships) that can successfully generate high employee motivation, so further studies of how such successes can be replicated for the rest of the US economy may be a promising direction for designing workforce policy.

How to Create Healthy Competition on the Job Market?

Just as firms have to compete on the product market with better, cheaper, goods, workers also have to compete on the job market, by investing in better education and making themselves more productive workers. As scholars have long pointed out, job market competition in the US may have failed to deliver either adequate education or adequate workforce productivity. Compared with their European counterparts, the American workforce lags notably behind in productivity; compared among themselves, massive numbers of American students fall behind their own compatriots in academic education, unable to complete either high school or college. As Goldin and Katz (2008) emphasized, the failure of the educational race is closely related to the highly inequitable distribution of economic opportunities in American society, caused by steep levels of poverty and income inequality, and addressing these social issues is a fundamental first step in the creation of healthier job market competition. If American policy-makers can indeed tackle these social problems, and succeed in creating an academic system that allows most students to aim and advance higher, will the nation close its productivity gap with Europe?

As findings from Chapter 4 suggest, better education does not necessarily translate into higher productivity on the job. In fact, job market competition in the US (and other Anglo-Saxon countries) has taken on the feature of a "rat race": job applicants keep running faster by investing in ever-higher academic credentials, even if the productivity of their actual job remains unchanged. The presence of "overeducation for the job" in the US economy implies that enabling disadvantaged students to advance further academically, albeit a vital task in itself, will not necessarily create healthier job market competition. In particular, although students may become less disadvantaged in getting *sorted* fairly on the job market (more diligent and proficient students get better jobs), jobs and applicants are still not *matched* (workers are more educated than their jobs need). Although both represent important dimensions of job market contests, *sorting* and *matching* are two very different types of competition. The former is an example of *rank-order tournament* (Lazear and Rosen 1981), where different firms use a *common* criterion (such as educational credential), and as a result, rank applicants the same way (higher-ranked applicants get better jobs); the latter is an example of Hotelling-style *spatial competition*, where, because of asset-specific needs, different firms have *different* criteria, and as a result, assess the same pool of applicants differently, some preferring the skill portfolio of one applicant while others preferring another (Gabsweiciz and Thisse 1986). In the language of Bar-Isaac et al. (2012), winners of rank-order tournaments are "popular" (wanted more or less by everyone), while winners of spatial competition carve out "niches" (preferred by some and disliked by others).

The distinction between these two forms of contests allows us to better understand the twin goals of healthy job market competition: it should not only produce "better workers" (in academic attainment), but also "better matches" (between workers and jobs). While helping left-behind students catch up in the classroom, as Goldin and Katz (2008) urge, is crucial in fulfilling the first goal, the second goal's fulfillment requires a strategy that takes seriously the spatial nature of competition: instead of raising the *level* of academic attainment by American students, this entails adjusting the *content* of academic education to make it easier for employers to assess the match with their own needs. As scholars of overeducation have long pointed out (Thurow 1976; Borghans and de Grip 2000a), once the content of education becomes an informative signal of productivity, employers no longer need to rely on "credential rankings," and mismatches in the form of overeducation will also disappear. To this end, the school-to-work literature (Kerckhoff 1995; Müller and Karle 1993; Müller and Shavit 1998) has identified three important pointers for reforms: (1) make the content of academic education more vocational and practical (i.e. *occupation-specificity*); (2) enhance the degree to which the quality of education meets the same standard nationwide (i.e. *standardization*); and (3) create more distinctive streaming of students into tracks towards tertiary versus vocational education at an early, secondary-education, stage (i.e. *stratification*). Reforms in these directions, plus the tackling of poverty and inequality that prevent many American students from fulfilling their academic potentials, may help create healthier job market competition on both fronts: first, a more educated workforce, and second, higher workforce productivity (as a result of better matches between jobs and workers). As Table 4.1 in Chapter 4 showed, none of the Anglo-Saxon countries score high on any of these three institutional features that may improve the matching between jobs and education. In other words, there is indeed much leeway for improvement in Anglo-Saxon economies.

At the same time, European economies may also benefit from some tinkering with their existing public education systems. As Chapter 4 showed, the workforce in these economies is less likely to experience overeducation for their job routine than their Anglo-Saxon counterparts. While this may be good news for routine tasks, this weaker collective incentive to invest in advanced academic education also puts the European workforce in a disadvantage when they face nonroutine tasks, such as radical innovation, which is an important part of the knowledge economy. In these countries, relatively high levels of *occupation-specificity*, *standardization*, and *stratification* in the education system have all reduced workers' incentives to use the "relative ranking" of academic degrees as productivity signals, discouraging them from aiming for higher academic attainment. Some downward adjustment on these features of the public education system will introduce some "noise" to the match between the "content" of education and work. In response, the job market will shift more weight to the "relative ranking" of education as productivity signals, raising workers' incentives to obtain more advanced education.

Of course, when the job match becomes noisier, the returns from investing deeply in job-specific human capital may also decline. As a result, it is important for European governments to balance the gain (i.e. encouraging academic education) from greater noise against the loss (i.e. discouraging skill investment). Nevertheless, assuming (realistically) that core institutions of coordinated capitalism do not change drastically, much of European economies' core institutions that sustain deep training (i.e. interfirm cooperation and firm–investor cooperation as Chapter 4 highlighted) will continue to do their job. As a result, coordinated economies will have some room for maneuver, allowing public policies to reorient towards academic education without compromising asset-specific training. In other words, the fact that coordination institutions do not change rapidly does not necessarily imply any deterministic constraint on the capacity for reform: in fact, it may generate a capacity for reform which is subtle enough to bypass some costly tradeoffs (such as between education and skills).

Social Policy

Innovation and the Politics of Redistribution

As Chapter 4 has shown, patterns of technological innovation in the economy directly affect patterns of income inequality in society. As a result, innovation will have interesting implications for how the society copes with inequality, not only through the partisan *politics* of *redistribution*, but also through the *institutions* of the *welfare state*. In this subsection I examine the implications for redistributive politics; in the next subsection I examine what kind of welfare state and social policy paradigm is the most effective institutional response to technological innovation.

Although Chapter 4's study of innovation and inequality focused on an outcome (i.e. wage inequality) that precedes the party politics of fiscal redistribution in taxes and welfare transfers, its explicit distinction between "inequality from the top" (p90/p50) and "inequality from the bottom" (p50/p10) may allow us to further understand how countries differ in their party competition over redistributive policies. As Lupu and Pontusson (2011) pointed out, the relative magnitudes of the top/median and median/bottom earnings gaps play an important role in the politics of fiscal redistribution. When "inequality from the top" is more prominent than "inequality from the bottom" (i.e. p90/p50 is larger than p50/p10, a condition the authors referred to as a *skew* in the inequality of income), the median voter will feel greater affinity with the poor, not only because of stronger perceptions of similar identities with them, but also because of deeper emersion in their social networks. As a result, the median voter's preference for fiscal redistribution increases, making the tax and welfare policy output of party competition more redistributive as well. When the book's findings on innovation and inequality are combined with Lupu and Pontusson's theoretical insight, we arrive at some interesting implications for redistributive politics. As Chapter 4 found, innovation in strategically coordinated capitalism will *narrow* the top/median wage gap but *widen* the median/bottom wage gap. As a result, a force for equality *before* taxes and transfers may also be a force for inequality *afterwards*: while innovation in strategically coordinated capitalism may alleviate the inequality of *wages* from the top, it may also indirectly exacerbate the inequality of *disposable income*, by weakening the median voter's preference for fiscal redistribution. In liberal market Anglo-Saxon economies, by contrast, the opposite dynamics for redistributive politics may unfold. Here, innovation widens the top/median wage gap but narrows the median/bottom wage gap. Median earners, as a result, are drawn further away from the top and closer to the bottom of society. This, according to Lupu and Pontusson, will intensify the median voter's preference for redistribution, and potentially reduce post-tax post-transfer inequality.

Depending on the type of capitalism, therefore, technological innovation may either reduce or strengthen the level of political support in society for fiscal redistribution in tax and welfare transfers. Since social democratic parties tend to be ideologically more sympathetic towards redistribution than their rightwing (especially non-Christian-democratic) counterparts, technological innovation may either undermine or consolidate the political constituency of social democratic parties. In European economies, innovation reduces the median voter's preference for fiscal redistribution, and as a result undercuts the redistributive message of center-left parties. In Anglo-Saxon economies, by contrast, innovation sharpens the median voter's preference for fiscal redistribution, which works to the political advantage of center-left parties. Historically, workers have had mixed feelings for the introduction of new technologies on the production floor, fearing, as the Luddites did, that they might be the first to be destroyed by the "gales of creative destruction" from innovation. This book's findings, however, imply that labor parties in some countries may embrace new technologies more enthusiastically than others, because they get to electorally benefit more from the economic impact of new technologies than labor parties in other countries. In particular, Anglo-Saxon social democratic parties may be more receptive to the prospect of technological innovation, a sentiment epitomized in Harold Wilson's (October 1, 1963) "white heat of technology" speech to the British Labour Party Annual Conference, in which he remarked that "the Britain that is going to be forged in the white heat of this (technological) revolution will be no place for restrictive practices or for outdated measures on either side of industry." Continental European social democrats, by comparison, may be more cautious in their embrace of new technological changes in the economy, which in turn may turn innovation into a favored theme for European center-right (especially liberal) parties.

Innovation and R&D, in other words, may be favored by the left in Anglo-Saxon countries and by the right in Europe. At this stage, this prediction is, of course, nothing more than a speculation. At least, it opens up the prospect for a theory about the *partisan politics of science policy*. Although technological innovation is an important component of comparative political economy (and in particular, the varieties of capitalism literature), innovation policy as an instrument of electoral competition has not been widely studied. While innovation policies are a key topic in the rich "systems of innovation" literature (Nelson 1993; Lundvall 2010; Edquist 1997; Edquist and McKelvey 2000), scholarship in this tradition has focused on designing innovation policies that can yield greater economic benefits for society, assuming for the state the classic task of maximizing social welfare as a benevolent planner. In reality, however, rulers of the state have more mixed incentives. The need for political survival requires that the policies they set maximize the welfare of their own "winning coalition" more than the rest of society (Bueno de Mesquita et al. 2003). As a result, on policies ranging from public education, corporate taxes to labor market training and social transfers, left and right parties often take very different positions. Although the statist literature (especially studies of the developmental state) has shed important lights on how the government may leverage its "first-among-equals" power over the society to encourage growth and innovation (Greene 2008; Trumbull 2004; Amable and Boyer

2001; Keller and Block 2013; Feldman and Francis 2003), it has not yet offered a sharp prediction about how the left and right may differ in their approach to innovation and why. This book's findings provide a potential framework in which such an inquiry can be pursued, so that government R&D, just like many other categories of public spending, can be studied as an element of partisan competition and electoral politics.

Is the Welfare State Dynamically Effective?

Besides insurance against risks, another important function of the welfare state is to offset, through redistribution, the poverty and inequality that often arise from economic and social transformation in modern societies. Some welfare states fulfill these poverty- and inequality-alleviating objectives far more effectively than others (Bradley et al. 2003; Allan and Scruggs 2004; Huber and Stephens 2001; Thelen 2014). Chapter 4's finding that new technologies create redistributive winners and losers highlights the importance of understanding welfare state effectiveness through the lens of innovation. In particular, is the welfare state *dynamically effective*? Here, the word "dynamically" implies that the welfare state should respond to the economy's *long-term dynamism* in the sense of Edith Penrose (1995), i.e. the capacity for innovation; the word "effective" implies that the welfare state should adequately *counter the potentially adverse income effects* of technological innovation, be it low income (poverty) or uneven distribution of income (inequality). Judged against this criterion, what kind of affluent capitalist economies have dynamically effective welfare states?

In liberal market Anglo-Saxon economies, as Chapter 4 found, technological innovation tends to be more radical. Radical innovation allows the high-income innovating sector to extract innovation rents from the rest of society, reducing the income of both median and low earners. Besides *lowering the income* of median and low earners, radical innovation also *widens the income gap* between the top and median, as the latter cedes markets and revenue to the former. A welfare state that is capable of adequately countering these adverse income outcomes (i.e. dynamically effective) should therefore target transfers not only at low but also median earners, which compensate both groups that were made worse off by innovation, as well as constrain inequality from the top. The strong emphasis on means-testing in Anglo-Saxon liberal welfare states is clearly inconsistent with this standard. Under means-testing, only the very bottom earners receive cash transfers from the state, and typically at a very low amount. As a result, Anglo-Saxon welfare states have little effect in closing the top/median income gap caused by technological innovation in their economies, and leave those made worse off by innovation inadequately compensated (low compensation for low earners and even less for median earners). Welfare states in Anglo-Saxon economies, in other words, are not dynamically effective.

Unlike liberal market Anglo-Saxon economies, coordinated European economies confront their social policy-makers with a different set of adverse income challenges from technological innovation. In these economies, as Chapter 4 found, innovation widens the median/bottom wage gap, causing the very low earners to lag further behind the rest of society. To the extent that low and median earners represent, respectively, the margin and core of the labor market, technological innovation in European economies widens the gap between labor market insiders and outsiders, reinforcing an insider/outsider divide that Rueda (2005, 2007) highlighted as an important problem for continental (but not Nordic) European countries. A welfare state that is capable of adequately addressing this challenge should target sufficient resources at the low-earning, vulnerable, margins of the labor market, and close the insider/outsider gap. The payroll-tax-funded earnings-related Christian democratic welfare state, however, has precisely the opposite effect. Since earnings-related benefits increase in the amount of earnings, such welfare state in essence devotes more resources to median than low earners, further widening the insider/outsider divide. Christian democratic welfare states, in other words, are not dynamically effective. Instead of closing the insider/outsider divide driven by technological innovation, they further consolidate the "dualism" (Thelen 2014) in continental European economies, which is itself a long-term consequence of many institutional features and policy choices, such as gender inequality, the skewing of employment protection in favor of prime-age male workers, and the lack of public sector services (Palier and Thelen 2010).

By contrast, Scandinavian social democratic welfare states have the opposite effect. Through income-tax-funded and often universal cash transfers, such welfare states not only redistribute more generous cash benefits to lower earners, but also provide a far more extensive set of public sector services such as healthcare, daycare, and labor market training, all of which help enhance gender equity and reintegrate outsiders in the labor market (Thelen 2014; Steinmo 2010; Gornick and Meyers 2003; Daly 2000). Furthermore, a high-quality public academic education system (Iversen and Stephens 2008; Nelson and Stephens 2012) helps prevent disadvantaged students from falling behind (which has become a serious problem for some other countries, as Goldin and Katz (2008) highlighted for the US). With fewer students left behind, fewer future earners will also lag behind the rest of society. Scandinavian welfare states, in other words, effectively offset the median/bottom income gap driven by technological innovation in their economies. Because they can adequately address the potentially adverse income impact of long-term economic dynamism (i.e. technological innovation), the Scandinavian welfare states are dynamically effective.

References

Abbott, Malcolm. 2000. "The Development of Vocational Education and Training in New Zealand." *Education Research and Perspectives* 27(1): 90–108.
Acemoglu, Daron. 2003a. "Patterns of Skill Premia." *Review of Economic Studies* 70(2): 199–230.
Acemoglu, Daron. 2003b. "Cross-Country Inequality Trends." *The Economic Journal* 113 (485): F121–49.
Acemoglu, Daron, and Jorn-Steffen Pischke. 1998. "Why Do Firms Train? Theory and Evidence." *Quarterly Journal of Economics* 113(1): 79–119.
Acemoglu, Daron, and James A. Robinson. 2005. *Economic Origins of Dictatorship and Democracy*. New York, NY: Cambridge University Press.
Acharya, Viral, Peter DeMarzo, and Ilan Kremer. 2011. "Endogenous Information Flows and the Clustering of Announcements." *American Economic Review* 101(7): 2955–79.
Afonso, Antonio, and Davide Furceri. 2010. "Government Size, Composition, Volatility, and Economic Growth." *European Journal of Political Economy* 26(4): 517–32.
Aghion, Philippe, Eve Caroli, and Cecilia García-Peñasola. 1999. "Inequality and Economic Growth: The Perspective of the New Growth Theories." *Journal of Economic Literature* 37(4): 1615–60.
Aghion, Philippe, Peter Howitt, and Giovanni Violante. 2002. "General Purpose Technology and Wage Inequality." *Journal of Economic Growth* 7(4): 315–42.
Aghion, Philippe, John van Reenen, and Luigi Zingales. 2013. "Innovation and Institutional Ownership." *American Economic Review* 103(1): 277–304.
Aizenman, Joshua, and Nancy Marion. 1993. "Policy Uncertainty, Persistence, and Growth." *Review of International Economics* 1(2): 145–63.
Akerlof, George A. 1970. "The Market for 'Lemons': Quality Uncertainty and the Market Mechanism." *Quarterly Journal of Economics* 84(3): 488–500.
Akerlof, George A. 1976. "The Economics of Caste and of the Rat Race and Other Woeful Tales." *Quarterly Journal of Economics* 90(4): 599–617.
Akerlof, George A. 1981. "Jobs as Dam Sites." *Review of Economic Studies* 48(1): 37–49.
Akkermans, Dirk, Carolina Castaldi, and Bart Los. 2009. "Do 'Liberal Market Economies' Really Innovate More Radically Than 'Coordinated Market Economies'?: Hall and Soskice Reconsidered." *Research Policy* 38(1): 181–91.
Albert, Michel. 1993. *Capitalism against Capitalism*. London: Whurt.
Alderson, Arthur, and François Nielsen. 2002. "Globalization and the Great U-Turn: Income Inequality Trends in 16 OECD Countries." *American Journal of Sociology* 107(5): 1244–99.

References

Allan, James, and Lyle Scruggs. 2004. "Political Partisanship and Welfare State Reform in Advanced Industrialized Societies." *American Journal of Political Science* 48(3): 496–512.

Allen, Franklin, and Douglas Gale. 2000. *Comparing Financial Systems*. Cambridge, MA: MIT Press.

Allmendinger, Jutta. 1989. "Educational Systems and Labor Market Outcomes." *European Sociological Review* 5(3): 231–50.

Alonso, Ricardo, Wouter Dessein, and Niko Matouschek. 2008. "When Does Coordination Require Centralization?" *American Economic Review* 98(1): 145–79.

Amable, Bruno. 2003. *The Diversity of Modern Capitalism*. Oxford: Oxford University Press.

Amable, Bruno, and Robert Boyer. 2001. "Europe's System(s) of Innovation." In *Technology and the Future of European Employment*, ed. Pascal Petit and Luc Soete, 425–51. Aldershot: Edward Elgar.

Andersen, Esben, and Bengt-Åke Lundvall. 1988. "Small National Systems of Innovation Facing Technological Revolutions: An Analytical Framework." In *Small Countries Facing the Technological Revolution*, ed. Christophe Freeman and Bengt-Åke Lundvall, 9–36. London: Pinter.

Ansell, Ben. 2008. "University Challenges: Explaining Institutional Change in Higher Education." *World Politics* 60(2): 189–230.

Ansell, Ben. 2010. *From the Ballot to the Blackboard: The Redistributive Political Economy of Education*. New York, NY: Cambridge University Press.

Antonucci, Tommaso, and Mario Pianta. 2002. "The Employment Effects of Product and Process Innovations in Europe." *International Review of Applied Economics* 16(3): 295–308.

Aoki, Masahiko. 1989. *Information, Incentives, and Bargaining in the Japanese Economy*. New York, NY: Cambridge University Press.

Aoki, Masahiko, Bo Gustafsson, and Oliver Williamson, ed. 1990. *The Firm as a Nexus of Treaties*. London: Sage.

Aoki, Masahiko, Gregory Jackson, and Hideaki Miyajima, ed. 2007. *Corporate Governance in Japan: Institutional Change and Organizational Diversity*. New York, NY: Oxford University Press.

Appelbaum, Eileen, Thomas Bailey, Peter Berg, and Arne Kalleberg. 2000. *Manufacturing Advantage: Why High-Performance Work Systems Pay Off*. Ithaca, NY: Cornell University Press.

Armingeon, Klaus, and Giuliano Bonoli, ed. 2006. *The Politics of Post-Industrial Welfare States: Adapting Post-War Social Policies to New Social Risks*. New York, NY: Routledge.

Arora, Ashish, and Andrea Fosfuri. 2002. "Licensing in the Chemical Industry." In *The Economics of Contracts: Theories and Applications*, ed. Eric Brousseau and Jean-Michel Glachant, 373–94. Cambridge: Cambridge University Press.

Arrow, Kenneth. 1974. *The Limits of Organization*. New York, NY: W. Norton.

Arundel, Anthony, Bengt-Åke Lundvall, Edward Lorenz, and Antoine Valeyre. 2007. "How Europe's Economies Learn: A Comparison of Work Organization and Innovate Mode for the EU-15." *Industrial and Corporate Change* 16(6): 680–93.

References

Asheim, Bjørn, and Arne Iskasen. 2002. "Regional Innovation Systems: The Integration of Local 'Sticky' and Global 'Ubiquitous' Knowledge." *Journal of Technology Transfer* 27(1): 77–86.

Autor, David, Lawrence Katz, and Alan Krueger. 1998. "Computing Inequality: Have Computers Changed the Labor Market?" *Quarterly Journal of Economics* 113(4): 1169–213.

Azoulay, Pierre. 2004. "Capturing Knowledge within and across Firm Boundaries: Evidence from Clinical Development." *American Economic Review* 94(5): 1591–612.

Baake, Pio, and Anette Boom. 2001. "Vertical Product Differentiation, Network Externalities, and Compatibility Decisions." *International Journal of Industrial Organization* 19(1–2): 267–84.

Baccaro, Lucio, and Marco Simoni. 2010. "Organizational Determinants of Wage Moderation." *World Politics* 62(4): 594–635.

Bacon, Frances. 1597. *Meditationes Sacrae*. London: John Windet.

Bagnasco, Arnaldo. 1977. *Le Tre Italie*. Bologna: Il Mulino.

Baldwin, Carliss. 1983. "Productivity and Labor Unions: An Application of the Theory of Self-Enforcing Contracts." *Journal of Business* 56(2): 155–85.

Baldwin, Kate, and John Huber. 2010. "Economic Versus Cultural Difference: Forms of Ethnic Diversity and Public Goods Provision." *American Political Science Review* 104(4): 644–62.

Bar-Isaac, Heski, Guillermo Caruana, and Vicente Cuñat. 2012. "Search, Design, and Market Structure." *American Economic Review* 102(2): 1140–60.

Barro, Robert, and Jong Wha Lee. 2013. "A New Data Set of Educational Attainment in the World: 1950–2010." *Journal of Development Economies* 104: 184–98.

Bartel, Anne, Casey Ichniowski, and Kathryn Shaw. 2007. "How Does Information Technology Affect Productivity? Plant-Level Comparisons of Product Innovation, Process Innovation, and Worker Skills." *Quarterly Journal of Economics* 122(4): 1721–58.

Bartels, Larry. 2008. *Unequal Democracy: The Political Economy of the New Gilded Age*. Princeton, NJ: Princeton University Press.

Bartling, Björn, Ernst Fehr, and Klaus M. Schmidt. 2012. "Screening, Competition, and Job Design: Economic Origins of Good Jobs." *American Economic Review* 102(2): 834–64.

Barzel, Yoram. 1997. *Economic Analysis of Property Rights*. New York, NY: Cambridge University Press.

Barzel, Yoram. 2002. *A Theory of the State: Economic Rights, Legal Rights, and the Scope of the State*. New York, NY: Cambridge University Press.

Battaglini, Marco, Salvatore Nunnari, and Thomas Palfrey. 2012. "Legislative Bargaining and the Dynamics of Public Investment." *American Political Science Review* 106(2): 407–29.

Battu, Harminder, and Peter Sloane. 2000. "Overeducation and Crowding Out in Britain." In *The Overeducated Worker? The Economics of Skill Utilization*, ed. Lex Borghans and Andries de Grip, 157–75. Northampton, MA: Edward Elgar.

Beck, Thorsten, George Clarke, Alberto Groff, Philip Keefer, and Patrick Walsh. 2001. "New Tools in Comparative Political Economy: The Database of Political Institutions." *World Bank Economic Review* 15(1): 165–76.

References

Beck, Thorsten, Aslı Demirgüç-Kunt, and Ross Levine. 2000. "A New Database on Financial Development and Structure." *World Bank Economic Review* 14: 597–605.

Becker, Gary. 1993. *Human Capital*. Chicago, IL: University of Chicago Press.

Béjar, Sergio, and Bumba Mukherjee. 2011. "Electoral Institutions and Growth Volatility: Theory and Evidence." *International Political Science Review* 32(4): 458–79.

Benavot, Aaron, and Cecilia Braslavsky, ed. 2007. *Social Knowledge in Comparative and Historical Perspective: Changing Curricula in Primary and Secondary Education*. Dordrecht: Springer.

Beramendi, Pablo. 2012. *The Political Geography of Inequality: Regions and Redistribution*. New York, NY: Cambridge University Press.

Berg, Gary. 2010. *Low Income Students and the Perpetuation of Inequality: Higher Education in America*. Burlington, VT: Ashgate.

Berggren, Christian. 1992. *Alternatives to Lean Production: Work Organization in the Swedish Auto Industry*. Ithaca, NY: ILR Press.

Berman, Eli, John Bound, and Stephen Machin. 1998. "Implications of Skill-Based Technological Change: International Evidence." *Quarterly Journal of Economics* 113(4): 1245–79.

Berry, Christopher. 2008. "Piling On: Multilevel Government and the Fiscal Common-Pool." *American Journal of Political Science* 52(4): 802–20.

Besanko, David, Ulrich Doraszelski, and Yaroslav Kryulov. 2014. "The Economics of Predation: What Drives Pricing When There Is Learning-by-Doing?" *American Economic Review* 104(3): 868–97.

Best, Michael. 1990. *New Competition: Institutions of Industrial Restructuring*. Cambridge: Polity Press.

Best, Michael. 2001. *The New Competitive Advantage: The Renewal of American Industry*. New York, NY: Oxford University Press.

Betts, Julian. 1997. "The Skill Bias of Technological Change in Canadian Manufacturing Industries." *The Review of Economics and Statistics* 79(1): 146–50.

Bhattacharya, Utpal, and Hazem Daouk. 2002. "The World Price of Insider Trading." *Journal of Finance* 57(1): 75–108.

Biggiero, Lucio. 2002. "The Location of Multinationals in Industrial Districts: Knowledge Transfer in Biomedicals." *Journal of Technology Transfer* 27(1): 111–22.

Black, Bernard, and Ronald Gilson. 1998. "Venture Capital and the Structure of Capital Markets: Banks Versus Stock Markets." *Journal of Financial Economics* 47(3): 243–77.

Board, Simon. 2011. "Relational Contracts and the Value of Loyalty." *American Economic Review* 101(7): 3349–67.

Boeri, Tito, Michael Burda, and Francis Kramarz, ed. 2008. *Working Hours and Job Sharing in the EU and USA: Are Europeans Lazy? Or Americans Crazy?* New York, NY: Oxford University Press.

Boix, Carles. 2003. *Democracy and Redistribution*. New York, NY: Cambridge University Press.

Boix, Carles. 2009. "The Conditional Relationship between Inequality and Development." *PS: Political Science & Politics* 42(4): 645–9.

Boix, Carles. 2010. "Origins and Persistence of Economic Inequality." *Annual Review of Political Science* 13: 489–516.

References

Bonoli, Giuliano. 1997. "Classifying Welfare States: A Two-Dimension Approach." *Journal of Social Policy* 26(3): 351–72.

Borghans, Lex, and Andries de Grip, ed. 2000a. *The Overeducated Worker? The Economics of Skill Utilization*. Northampton, MA: Edward Elgar.

Borghans, Lex, and Andries de Grip. 2000b. "The Debate in Economics about Skill Utilization." In *The Overeducated Worker? The Economics of Skill Utilization*, ed. Lex Borghans and Andries de Grip, 3–26. Northampton, MA: Edward Elgar.

Bowles, Samuel, and Herbert Gintis. 1990. "Contested Exchange: New Microfoundations of the Political Economy of Capitalism." *Politics and Society* 18(2): 165–222.

Bradley, David, Evelyne Huber, Stephanie Moller, François Nielsen, and John D. Stephens. 2003. "Distribution and Redistribution in Postindustrial Democracies." *World Politics* 55 (2): 193–228.

Bradley, David, and John D. Stephens. 2007. "Employment Performance in OECD Countries: A Test of Neoliberal and Institutional Hypotheses." *Comparative Political Studies* 40(12): 1486–510.

Braunthal, Gerard. 1965. *The Federation of German Industry in Politics*. Ithaca, NY: Cornell University Press.

Broberg, Andreas, Holger Wessels, Michael Lamb, and C. Philip Hwang. 1997. "Effects of Day Care on the Development of Cognitive Abilities in 8-Year-Olds: A Longitudinal Study." *Developmental Psychology* 33(1): 62–9.

Brouwer, Erik, Alfred Kleinknecht, and Jeroen Reijnen. 1993. "Employment Growth and Innovation at the Firm Level." *Journal of Evolutionary Economics* 3(2): 153–9.

Brunetti, Aymo. 1998. "Policy Volatility and Economic Growth: A Comparative, Empirical Analysis." *European Journal of Political Economy* 14(1): 35–52.

Bueno de Mesquita, Bruce, Alastair Smith, Randolph M. Siverson, and James D. Morrow. 2003. *The Logic of Political Survival*. Cambridge, MA: MIT Press.

Busch, Marc. 1999. *Trade Warriors: States, Firms, and Strategic Trade Policy in High-Technology Competition*. New York, NY: Cambridge University Press.

Busemeyer, Marius, and Carsten Jensen. 2012. "The Impact of Economic Coordination and Educational Institutions on Individual-Level Preferences for Academic and Vocational Education." *Socio-Economic Review* 10(3): 525–47.

Camerer, Colin. 1988. "Gifts as Economic Signals and Social Symbols." *American Journal of Sociology* 94 (*Supplement: Organizations and Institutions: Sociological and Economic Approaches to the Analysis of Social Structure*): S180–214.

Cameron, David. 1978. "The Expansion of the Public Economy: A Comparative Analysis." *American Political Science Review* 72(4): 1243–61.

Cameron, David. 1984. "Social Democracy, Corporatism, Labor Quiescence, and the Representation of Economic Interests in Advanced Capitalist Society." In *Order and Conflict in Contemporary Capitalism: Studies in the Political Economy of Western European Nations*, ed. John Goldthorpe, 143–78. Oxford: Oxford University Press.

Campbell, Frances, Elizabeth Pungello, Shari Miller-Johnson, Margaret Burchinal, and Craig Ramey. 2001. "The Development of Cognitive and Academic Abilities: Growth Curves from an Early Childhood Educational Experiment." *Developmental Psychology* 37(2): 231–41.

References

Carmichael, H. Lorne, and Bentley MacLeod. 1997. "Gift Giving and the Evolution of Cooperation." *International Economic Review* 38(3): 485–509.

Casper, Steven. 2000. "Institutional Adaptiveness, Technology Policy, and the Diffusion of New Business Models: The Case of German Biotechnology." *Organizational Study* 21(5): 887–914.

Casper, Steven. 2009. "Can New Technology Firm Succeed in Coordinated Market Economies? A Response to Herrmann and Lange." *Socio-Economic Review* 7(2): 209–15.

Casper, Steven, and Richard Whitley. 2004. "Managing Competences in Entrepreneurial Technology Firms: A Comparative Institutional Analysis of Germany, Sweden, and the UK." *Research Policy* 33(1): 89–106.

Casper, Steven, Rogers Hollingsworth, and Richard Whitley. 2005. "Varieties of Capitalism: Comparative Institutional Approaches to Economic Organization and Innovation." In *Innovation and Institutions: A Multidisciplinary Review of the Study of Innovation Systems*, ed. Steven Casper and Frans van Waarden, 193–228. Cheltenham, UK: Edward Elgar.

Castells, Manuel. 1996. *The Rise of the Network Society*. Cambridge, MA: Blackwell Publishers.

Castles, Frances. 1985. *Working Class and Welfare: Reflections on the Political Development of the Welfare State in Australia and New Zealand*. Wellington: Allen & Unwin.

Castles, Frances, and Steve Dowrick. 1990. "The Impact of Government Spending Levels on Medium-Term Economic Growth in the OECD, 1960–85." *Journal of Theoretical Politics* 2(2): 173–204.

Chamberlin, Edward. 1937. "Monopolistic or Imperfect Competition?" *Quarterly Journal of Economics* 51(4): 557–80.

Chandler, Alfred. 1977. *The Visible Hand: The Managerial Revolution in American Business*. Cambridge, MA: Harvard University Press.

Chandler, Alfred. 1990. *Scale and Scope: The Dynamics of Industrial Capitalism*. Cambridge, MA: Harvard University Press.

Chaney, Paul, Timothy Devinney, and Russell Winter. 1991. "The Impact of New Product Introductions on the Market Value of Firms." *Journal of Business* 64(4): 573–610.

Chang, Eric, Mark Keyser, Drew Linzer, and Ronald Rogowski. 2011. *Electoral Systems and the Balance of Consumer-Producer Power*. New York, NY: Cambridge University Press.

Chassang, Sylvain. 2010. "Building Routines: Learning, Cooperation, and the Dynamics of Incomplete Relational Contracts." *American Economic Review* 100(1): 448–65.

Chen, Sheng-Syan, Kim Wai Ho, and Kueh Hwa Ik. 2005. "The Wealth Effect of New Product Introductions on Industry Rivals." *Journal of Business* 78(3): 969–96.

Clarke, Kevin, and Randall Stone. 2008. "Democracy and the Logic of Political Survival." *American Political Science Review* 102(3): 387–92.

Clasen, Jochen, and Daniel Clegg, ed. 2011. *Regulating the Risk of Unemployment: National Adaptations to Postindustrial Labor Markets in Europe*. New York, NY: Oxford University Press.

Coase, Ronald. 1937. "The Nature of the Firm." *Econometrica* 4(16): 386–405.

References

Coase, Ronald. 1960. "The Problem of Social Cost." *Journal of Law and Economics* 3(1): 1–44.

Coleman, William, and Wyn Grant. 1988. "The Organizational Cohesion and Political Access of Business: A Study of Comprehensive Associations." *European Journal of Political Research* 16(5): 467–87.

Comanor, William. 1967. "Market Structure, Product Differentiation, and Industrial Research." *Quarterly Journal of Economics* 81(4): 639–57.

Cosci, Stefania. 1993. *Credit Rationing and Asymmetric Information*. Brookfield, VT: Dartmouth.

Crouch, Colin. 1993. *Industrial Relations and European State Traditions*. Oxford: Clarendon Press.

Crouch, Colin, David Finegold, and Mari Sako. 1999. *Are Skills the Answer? The Political Economy of Skill Creation in Advanced Industrial Countries*. Oxford: Oxford University Press.

Crouch, Colin, Martin Shröder, and Helmut Voelzkow. 2009. "Regional and Sectoral Varieties of Capitalism." *Economy and Society* 38(4): 654–78.

Crouch, Colin, and Wolfgang Streeck. 1997. "Introduction: The Future of Capitalist Diversity." In *Political Economy of Modern Capitalism*, ed. Colin Crouch and Wolfgang Streeck, 1–18. London: Sage.

Culpepper, Pepper. 2003. *Creating Cooperation: How States Develop Human Capital in Europe*. Ithaca, NY: Cornell University Press.

Culpepper, Pepper. 2005. "Institutional Change in Contemporary Capitalism: Coordinated Financial Systems since 1990." *World Politics* 57(2): 173–99.

Culpepper, Pepper. 2011. *Quiet Politics and Business Power: Corporate Control in Europe and Japan*. New York, NY: Cambridge University Press.

Cusack, Thomas, Torben Iversen, and David Soskice. 2010. "Co-evolution of Capitalism and Political Representation: The Choice of Electoral Systems." *American Political Science Review* 104(2): 393–403.

Daly, Mary. 2000. "A Fine Balance: Women's Labor Market Participation in International Comparison." In *Welfare and Work in the Open Economy. Volume II. Diverse Responses to Common Challenges*, ed. Fritz Scharpf and Vivien Schmidt, 467–510. New York, NY: Oxford University Press.

Darby, Michael R., and Edi Karni. 1973. "Free Competition and the Optimal Amount of Fraud." *Journal of Law and Economics* 16(1): 67–88.

de Meza, David, and David Webb. 1990. "Risk, Adverse Selection and Capital Market Failure." *The Economic Journal* 100(399): 206–14.

Deeg, Richard. 2009. "The Rise of Internal Capitalist Diversity? Changing Patterns of Finance and Corporate Governance in Europe." *Economy and Society* 38(4): 552–79.

Demsetz, Harold. 1988. *Ownership, Control, and the Firm: The Organization of Economic Activity*. Oxford: Blackwell.

Desai, Raj, Anders Olofsgård, and Tarik Yousef. 2003. "Democracy, Inequality, and Inflation." *American Political Science Review* 97(3): 391–406.

Dietl, Helmut. 1998. *Capital Markets and Corporate Governance in Japan, Germany and the United States: Organizational Response to Market Inefficiencies*. London: Routledge.

References

DiPrete, Thomas. 2002. "Life Course Risks, Mobility Regimes, and Mobility Consequences: A Comparison of Sweden, Germany, and the United States." *American Journal of Sociology* 108(2): 267–309.

Dixit, Avinash, and Joseph Stiglitz. 1977. "Monopolistic Competition and Optimal Product Diversity." *American Economic Review* 67(3): 297–308.

Djelic, Marie-Laure. 1998. *Exporting the American Model: The Postwar Transformation of European Business*. Oxford: Oxford University Press.

Dollar, David. 1992. "Outward-Oriented Developing Economies Really Do Grow More Rapidly: Evidence from 95 LDCs, 1976–1985." *Economic Development and Cultural Change* 40(3): 523–44.

Dollar, David, and Edward Wolff. 1993. *Competitiveness, Convergence, and International Specialization*. Cambridge, MA: MIT Press.

Doucouliagos, Hristos, and Mehmet Ali Ulubaşoğlu. 2008. "Democracy and Economic Growth: A Meta-Analysis." *American Journal of Political Science* 52(1): 61–83.

Doukas, John, and Lorne Switzer. 1992. "The Stock Market's Valuation of R&D Spending and Market Concentration." *Journal of Economics and Business* 44(2): 95–114.

Dow, Gregory. 2003. *Governing the Firm: Workers' Control in Theory and Practice*. New York, NY: Cambridge University Press.

Ebbinghaus, Bernhard. 2006. *Reforming Early Retirement in Europe, Japan, and the USA*. New York, NY: Oxford University Press.

Edlin, Aaron, and Joseph Stiglitz. 1995. "Discouraging Rivals: Managerial Rent-Seeking and Economic Inefficiencies." *American Economic Review* 85(5): 1301–12.

Edquist, Charles, ed. 1997. *Systems of Innovation: Technologies, Institutions, and Organizations*. London: Pinter/Casell Academic.

Edquist, Charles, Leif Hommen, and Maureen McKelvey. 2001. *Innovation and Employment: Process versus Product Innovation*. Northampton, MA: Edward Elgar.

Edquist, Charles, and Maureen McKelvey, ed. 2000. *Systems of Innovation: Growth, Competitiveness, and Employment*. Cheltenham: Edward Elgar.

Ellingsen, Tore, and Magnus Johannesson. 2008. "Pride and Prejudice: The Human Side of Incentive Theory." *American Economic Review* 93(3): 990–1008.

Engle, Robert. 1982. "Autoregressive Conditional Heteroskedasticity with Estimates of the Variance of United Kingdom Inflation." *Econometrica* 50(4): 987–1007.

Esping-Andersen, Gøsta. 1985. *Politics against Markets: The Social Democratic Road to Power*. Princeton: Princeton University Press.

Esping-Andersen, Gøsta. 1990. *Three Worlds of Welfare Capitalism*. Princeton: Princeton University Press.

Esping-Andersen, Gøsta. 1999. *Social Foundations of Postindustrial Economies*. New York, NY: Oxford University Press.

Estévez-Abe, Margarita. 2006. "Gendering the Varieties of Capitalism: A Study of Occupational Segregation by Sex in Advanced Industrialized Societies." *World Politics* 59(1): 142–75.

Estevez-Abe, Margarita, Torben Iversen, and David Soskice. 2001. "Social Protection and the Formation of Skills: A Reinterpretation of the Welfare State." In *Varieties of Capitalism: The Institutional Foundations of Comparative Advantage*, ed. Peter Hall and David Soskice, 145–83. New York, NY: Oxford University Press.

References

European Private Equity and Venture Capital Association. 2011. *2011 Pan-European Private Equity Performance Benchmarks Study*. Brussels.

Evangelista, Rinaldo. 1999. *Knowledge and Investment: The Sources of Innovation in Industry*. Cheltenham: Edward Elgar.

Falk, Armin, and Michael Kosfeld. 2006. "The Hidden Costs of Control." *American Economic Review* 96(5): 1611–30.

Fama, Eugene. 1970. "Efficient Capital Markets: A Review of Theory and Empirical Work." *Journal of Finance* 25(2): 383–417.

Farrell, Henry. 2009. *The Political Economy of Trust: Institutions, Interests, and Inter-Firm Cooperation in Italy and Germany*. New York, NY: Cambridge University Press.

Fehr, Ernst, Holger Herz, and Tom Wilkening. 2013. "The Lure of Authority: Motivation and Incentive Effects of Power." *American Economic Review* 103(4): 1325–59.

Fehr, Ernst, Erich Kirchler, Andreas Weichbold, and Simon Gächter. 1998. "When Social Norms Overpower Competition: Gift Exchange in Experimental Labor Markets." *Journal of Labor Economics* 16(2): 324–51.

Fehr, Ernst, and Bettina Rockenbach. 2003. "Detrimental Effects of Sanctions on Human Altruism." *Nature* 422: 137–40.

Feldman, Maryann, and Johanna Francis. 2003. "Fortune Favors the Prepared Region: The Case of Entrepreneurship and the Capitol Region Biotechnology Cluster." *European Planning Studies* 11(7): 765–88.

Financial Stability Board and International Monetary Fund. 2009. *Financial Crisis and Information Gaps: Report to the G20 Finance Ministers and Central Bank Governors*. Washington, DC.

Finegold, David, and Karin Wagner. 1999. "The German Skill-Creation System and Team-Based Production: Competitive Asset or Liability?" In *The German Skills Machine: Sustaining Comparative Advantage in a Global Economy*, ed. Pepper Culpepper and David Finegold, 115–58. New York, NY: Berghahn Books.

Fligstein, Neil. 2001. *The Architecture of Markets: An Economic Sociology of Twenty-First-Century Capitalist Societies*. Princeton, NJ: Princeton University Press.

Fransman, Martin. 1990. *The Market and Beyond: Cooperation and Competition in Information Technology Development in the Japanese System*. Cambridge: Cambridge University Press.

Franzese, Robert J. 2002. *Macroeconomic Policies of Developed Democracies*. New York, NY: Cambridge University Press.

Freeman, Christopher, and Luc Soete. 1997. *The Economics of Industrial Innovation*. London: Pinter.

Frey, Bruno. 1997a. *Not Just for the Money: An Economic Theory of Personal Motivation*. Cheltenham: Edward Elgar.

Frey, Bruno. 1997b. "On the Relationship between Intrinsic and Extrinsic Work Motivation." *International Journal of Industrial Organization* 15(4): 427–39.

Gabsweiciz, Jean Jaskold, and Jacques-François Thisse. 1986. "On the Nature of Competition with Differentiated Products." *The Economic Journal* 96(381): 160–72.

Galbraith, James. 1998. *Created Unequal: The Crisis in American Pay*. New York, NY: Free Press.

Galbraith, James, and Maureen Berner, ed. 2001. *Inequality and Industrial Change: A Global View*. New York, NY: Cambridge University Press.

References

Galbraith, James, and Hyunsub Kum. 2005. "Estimating the Inequality of Household Incomes: A Statistical Approach to the Creation of a Dense and Consistent Global Data Set." *Review of Income and Wealth* 51(1): 115–43.

Gambardella, Alfonso. 1995. *Science and Innovation: The US Pharmaceutical Industry during the 1980s*. Cambridge: Cambridge University Press.

Gangl, Markus. 2004. "Welfare States and the Scar Effects of Unemployment: A Comparative Analysis of the United States and West Germany." *American Journal of Sociology* 109(6): 1319–64.

Garrett, Geoffrey. 1998. *Partisan Politics in the Global Economy*. New York, NY: Cambridge University Press.

Gershkov, Alex, and Benny Modovanu. 2009. "Learning about the Future and Dynamic Efficiency." *American Economic Review* 99(4): 1576–87.

Gilbert, Richard, and David Newbery. 1982. "Preemptive Patenting and the Persistence of Monopoly." *American Economic Review* 72(3): 514–26.

Ginsburg, Douglas. 2005. "Comparing Antitrust Enforcement in the United States and Europe." *Journal of Competition Law and Economics* 1(3): 427–39.

Gnosh, Arghya, and Hodaka Morita. 2008. "An Economic Analysis of Platform Sharing." *Journal of the Japanese and International Economies* 22(2): 164–86.

Golden, Mariam. 1997. *Heroic Defeats: The Politics of Job Loss*. New York, NY: Cambridge University Press.

Goldfarb, Avi, and Mo Xiao. 2011. "Who Thinks About the Competition? Managerial Ability and Strategic Entry in US Local Telephone Markets." *American Economic Review* 101(7): 3130–61.

Goldin, Claudia, and Lawrence Katz. 2008. *The Race between Education and Technology*. Cambridge: Harvard University Press.

Gompers, Paul. 1995. "Optimal Investment, Monitoring, and the Staging of Venture Capital." *Journal of Finance* 50(5): 1461–89.

Gornick, Janet, and Marcia Meyers. 2003. *Families that Work: Policies for Reconciling Parenthood and Employment*. New York, NY: Sage.

Gourevitch, Peter, and James Shinn. 2005. *Political Power and Corporate Control: The New Global Politics of Corporate Governance*. Princeton, NJ: Princeton University Press.

Graf, Holger. 2006. *Networks in the Innovation Process: Local and Regional Interactions*. Cheltenham: Edward Elgar.

Graf, Lukas. 2008. "Applying the Varieties of Capitalism Approach to Higher Education: A Case Study of the Internationalization Strategies of German and British Universities." Wissenschaftszentrum Berlin für Sozialforschung (WZB) Discussion Paper No. SP I 2008-507.

Green, Francis, David Ashton, and Brendan Burchell. 2000. "Are British Workers Becoming More Skilled?" In *The Overeducated Worker? The Economics of Skill Utilization*, ed. Lex Borghans and Andries de Grip, 77–109. Northampton, MA: Edward Elgar.

Greenan, Nathalie, Yannick L'Horty, and Jacques Mairesse, ed. 2002. *Productivity, Inequality, and the Digital Economy: A Transatlantic Perspective*. Cambridge, MA: MIT Press.

Greene, Megan. 2008. *The Origins of the Developmental State in Taiwan: Science Policy and the Quest for Modernization*. Cambridge, MA: Harvard University Press.

References

Greif, Avner. 2006. *Institutions and the Path to the Modern Economy: Lessons from Medieval Trade*. Cambridge: Cambridge University Press.

Griliches, Zvi. 2000. *R&D, Education, and Productivity: A Retrospective*. Cambridge, MA: Harvard University Press.

Grossman, Sanford, and Oliver Hart. 1986. "The Costs and Benefits of Ownership: A Theory of Vertical and Lateral Integration." *Journal of Political Economy* 94(4): 691–719.

Hacker, Jacob S. 2003. "Call It the Family Risk Factor." *New York Times* 11 January A15. <http://www.nytimes.com/2004/01/11/opinion/call-it-the-family-risk-factor.html>.

Hacker, Jacob S. 2004. "Privatizing Risk without Privatizing the Welfare State: The Hidden Politics of Welfare State Retrenchment in the United States." *American Political Science Review* 98(2): 243–60.

Hall, Peter, and Daniel Gingerich. 2004. "Varieties of Capitalism and Institutional Complementarities in the Political Economy: An Empirical Analysis." Max-Planck-Institut für Gesellschaftsforschung Discussion Paper 04/5.

Hall, Peter, and Daniel Gingerich. 2009. "Varieties of Capitalism and Institutional Complementarities in the Political Economy: An Empirical Analysis." *British Journal of Political Science* 39(3): 449–82.

Hall, Peter, and David Soskice. 2001. "An Introduction to Varieties of Capitalism." In *Varieties of Capitalism: The Institutional Foundations of Comparative Advantage*, ed. Peter Hall and David Soskice, 1–70. New York, NY: Oxford University Press.

Hall, Peter, and Kathleen Thelen. 2009. "Institutional Change in Varieties of Capitalism." *Socio-Economic Review* 7(1): 7–34.

Hallerberg, Mark, and Patrik Marier. 2004. "Executive Authority, the Personal Vote, and Budget Discipline in Latin American and Caribbean Countries." *American Journal of Political Science* 48(3): 571–87.

Hancké, Bob, Martin Rhodes, and Mark Thatcher, ed. 2007. *Beyond Varieties of Capitalism: Conflict, Contradictions, and Complementarities in the European Economy*. Oxford: Oxford University Press.

Hannan, Damian, David Raffe, and Emer Smyth. 1996. "Cross-National Research on School to Work Transitions: An Analytic Framework." In *Youth Transitions in Europe: Theories and Evidence. Third ESF Workshop of the Network on Transitions in Youth. La Ciotat, September 1996*, ed. Patrick Werquin, Richard Breen, and Jordi Planas, 409–42. Marseille: CEREQ.

Harris, Milton, and Artur Raviv. 1993. "Differences of Opinion Make a Horse Race." *Review of Financial Studies* 6(3): 473–506.

Harrison, J. Michael, and David Kreps. 1978. "Speculative Investor Behavior in a Stock Market with Heterogeneous Expectations." *Quarterly Journal of Economics* 92(2): 323–36.

Hart, Oliver. 1995. *Firms, Contracts, and Financial Structure*. New York, NY: Oxford University Press.

Hart, Oliver. 2002. "Norms and the Theory of the Firm." In *The Economics of Contracts: Theories and Applications*, ed. Eric Brousseau and Jean-Michel Glachant, 180–92. Cambridge: Cambridge University Press.

Harvey, David. 2010. *The Enigma of Capital and the Crises of Capitalism*. New York, NY: Oxford University Press.

References

Hassel, Anke. 2006. *Wage Setting, Social Pacts, and the Euro: A New Role for the State.* Amsterdam: University of Amsterdam Press.

Hatzichronoglou, Thomas. 1997. "Revision of the High-Technology Sector and Product Classification." OECD Science, Technology and Industry Working Papers.

Heckman, James. 2000. "*Policies to Foster Human Capital.*" Research in Economics 54(1): 3–56.

Hellman, Thomas, Laura Lindsey, and Manju Puri. 2008. "Building Relationships Early: Banks in Venture Capital." Review of Financial Studies 21(2): 513–41.

Helpman, Elhanan. 1998. *General Purpose Technologies and Economic Growth.* Cambridge, MA: MIT Press.

Herrigel, Gary. 1996. *Industrial Constructions: The Sources of German Industrial Power.* New York, NY: Cambridge University Press.

Herrigel, Gary, and Charles Sabel. 1999. "Craft Production in Crisis: Industrial Restructuring in Germany during the 1990s." In *The German Skills Machine: Sustaining Comparative Advantage in a Global Economy*, ed. Pepper Culpepper and David Finegold, 77–114. New York, NY: Berghahn Books.

Herrmann, Andrea. 2008. "Rethinking the Link between Labor Market Flexibility and Corporate Competitiveness: A Critique of the Institutionalist Literature." Socio-Economic Review 6(4): 637–69.

Hicken, Allen, Shanker Satyanath, and Ernest Sergenti. 2005. "Political Institutions and Economic Performance." American Journal of Political Science 49(4): 897–907.

Hicks, Alex, and Lane Kenworthy. 1998. "Cooperation and Political Economic Performance in Affluent Democratic Capitalism." American Journal of Sociology 103(6): 1631–72.

Hollingsworth, J. Rogers. 1993. "Variation among Nations in the Logic of Manufacturing Sectors and International Competitiveness." In *Technology and the Wealth of Nations: The Dynamics of Constructed Advantage*, ed. Dominique Foray and Christopher Freeman, 301–21. New York, NY: Pinter.

Hollingsworth, J. Rogers, and Boyer, Robert, ed. 1997. *Contemporary Capitalism: The Embeddedness of Institutions.* New York, NY: Cambridge University Press.

Holmström, Bengt. 1999. "The Firm as a Subeconomy." Journal of Law, Economics, and Organization 51(1): 74–102.

Hotelling, Harold. 1929. "Stability in Competition." The Economic Journal 39(153): 41–57.

Howell, Chris. 2005. *Trade Unions and the State: Constructing Industrial Relations Institutions in Britain, 1890–2000.* Princeton, NJ: Princeton University Press.

Huber, Evelyne, Charles Ragin, John D. Stephens, David Brady, and Jason Beckfield. 2004. Comparative Welfare States Data Set, Northwestern University, University of North Carolina, Duke University and Indiana University.

Huber, Evelyne, and John D. Stephens. 2001. *Development and Crisis of the Welfare State: Parties and Policies in Global Markets.* Chicago, IL: University of Chicago Press.

Huo, Jingjing. 2009. *Third Way Reforms: Social Democracy after the Golden Age.* New York, NY: Cambridge University Press.

Huo, Jingjing, Moira Nelson, and John D. Stephens. 2008. "Decommodification and Activation in Social Democratic Policy: Resolving the Paradox." Journal of European Social Policy 18(5): 1–20.

References

Huo, Jingjing, and John D. Stephens. 2015. "From Industrial Corporatism to the Social Investment State." In *Oxford Handbook on Transformation of the State*, ed. Stephan Leibfried, Frank Nullmeier, Evelyne Huber, Matthew Lange, Jonah Levy, and John D. Stephens. New York, NY: Oxford University Press.

Ichniowski, Casey, and Kathryn Shaw. 2003. "Beyond Incentive Pay: Insiders' Estimates of the Value of Complementary Human Resource Management Practices." *Journal of Economic Perspectives* 17(1): 155–80.

Iversen, Torben. 2005. *Capitalism, Democracy, and Welfare*. New York, NY: Cambridge University Press.

Iversen, Torben, Jonas Pontusson, and David Soskice, ed. 2000. *Unions, Employers, and Central Banks: Macroeconomic Coordination and Institutional Change in Social Market Economies*. New York, NY: Cambridge University Press.

Iversen, Torben, and Frances Rosenbluth. 2010. *Women, Work, and Politics: The Political Economy of Gender Inequality*. New Haven, CT: Yale University Press.

Iversen, Torben, and David Soskice. 2006. "Electoral Institutions and the Politics of Coalitions: Why Some Democracies Distribute More than Others." *American Political Science Review* 100(2): 165–81.

Iversen, Torben, and David Soskice. 2009. "Distribution and Redistribution: The Shadow of the Nineteenth Century." *World Politics* 61(3): 438–86.

Iversen, Torben, and David Soskice. 2010. "Real Exchange Rates and Competitiveness: The Political Economy of Skill Formation, Wage Compression, and Electoral Systems." *American Political Science Review* 104(3): 601–23.

Iversen, Torben, and John D. Stephens. 2008. "Partisan Politics, the Welfare State, and Three World of Human Capital Formation." *Comparative Political Studies* 41(4): 600–37.

Iversen, Torben, and Anne Wren. 1998. "Equality, Employment and Budgetary restraint: The Trilemma of the Service Economy." *World Politics* 50(4): 507–46.

Jackson, Gregory. 2009. "The Japanese Firm and Its Diversity." *Economy and Society* 38(4): 606–29.

Jaffee, Dwight, and Thomas Russell. 1976. "Imperfect Information, Uncertainty and Credit Rationing." *Quarterly Journal of Economics* 90(4): 651–66.

Jenson, Jane. 2011. "Redesigning Citizenship Regimes after Neoliberalism: Moving towards Social Investment." In *Towards a Social Investment Welfare State? Ideas, Policies and Challenges*, ed. Nathalie Morel, Bruno Palier, and Joakim Palme, 61–90. Bristol: Polity Press.

Johnson, Joel W., and Jessica S. Wallack. 2012. "Electoral Systems and the Personal Vote." <http://hdl.handle.net/1902.1/17901>.

Johnson, Justin, and David Myatt. 2006. "On the Simple Economics of Advertising, Marketing, and Product Design." *American Economic Review* 96(3): 756–84.

Jones, Erik. 2008. *Economic Adjustment and Political Transformation in Small States*. New York, NY: Oxford University Press.

Jorgenson, Dale. 1995. *Productivity: Postwar US Economic Growth*. Cambridge, MA: MIT Press.

Kam, Cindy, and Robert J. Franzese. 2007. *Modeling and Interpreting Interactive Hypotheses in Regression Analysis*. Ann Arbor: University of Michigan Press.

References

Kamenica, Emir, and Matthew Gentzkow. 2011. "Bayesian Persuasion." *American Economic Review* 101(6): 2590–615.

Kaplan, Steven, and Per Strömberg. 2001. "Venture Capitalists as Principals: Contracting, Screening, and Monitoring." *American Economic Review* 91(2): 426–30.

Katz, Lawrence, and Kevin Murphy. 1992. "Changes in Relative Wages, 1963–1987: Supply and Demand Factors." *Quarterly Journal of Economics* 107(1): 35–78.

Katzenstein, Peter. 1984. *Corporatism and Change: Austria, Switzerland, and the Politics of Industry*. Ithaca, NY: Cornell University Press.

Katzenstein, Peter. 1985. *Small States in World Markets: Industrial Policy in Europe*. Ithaca, NY: Cornell University Press.

Kaya, Ayça, and Galina Vereschagina. 2014. "Partnerships versus Corporations: Moral Hazard, Sorting, and Ownership Structure." *American Economic Review* 104(1): 291–307.

Kayser, Mark Andreas. 2005. "Who Surfs, Who Manipulates? The Determinants of Opportunistic Election Timing and Electorally Motivated Economic Intervention." *American Political Science Review* 99(1): 17–27.

Keefer, Philip, and David Stasavage. 2003. "The Limits of Delegation: Veto Players, Central Bank Independence, and the Credibility of Monetary Policy." *American Political Science Review* 97(3): 407–23.

Keller, Matthew, and Fred Block. 2013. "Explaining the Transformation in the US Innovation System: The Impact of a Small Government Program." *Socio-Economic Review* 11(4): 629–56.

Kelly, Nathan. 2005. "Political Choice, Public Policy, and Distributional Outcomes." *American Journal of Political Science* 49(4): 865–80.

Kelly, Nathan, and Peter Enns. 2010. "Inequality and the Dynamics of Public Opinion: The Self-Reinforcing Link between Inequality and Mass Preferences." *American Journal of Political Science* 54(4): 855–70.

Kelsey, Jane. 1993. *Rolling Back the State: Privatization of Power in Aotearoa/New Zealand*. Wellington: Bridge William Books.

Kenney, Martin, and Richard Florida. 1993. *Beyond Mass Production: The Japanese System and Its Transfer to the US*. New York, NY: Oxford University Press.

Kenworthy, Lane. 2001. "Wage-Setting Measures: A Survey and Assessment." *World Politics* 54(1): 57–98.

Kenworthy, Lane. 2003. "Quantitative Indicators of Corporatism: A Survey and Assessment." *International Journal of Sociology* 33(3): 10–44.

Kenworthy, Lane. 2004. *Egalitarian Capitalism: Jobs, Incomes, and Growth in Affluent Countries*. New York, NY: Sage.

Kenworthy, Lane. 2008. *Jobs with Equality*. New York, NY: Oxford University Press.

Kerckhoff, Alan. 1995. "Institutional Arrangements and Stratification Processes in Industrial Societies." *Annual Review of Sociology* 21: 323–47.

Kerr, Clark, John Dunlop, Frederick Harbison, and Charles Myers. 1960. *Industrialism and Industrial Man*. New York, NY: Oxford University Press.

Kim, So Young. 2007. "Openness, External Risk, and Volatility: Implications for the Compensation Hypothesis." *International Organization* 61(1): 181–216.

King, Robert, and Ross Levine. 1993. "Finance and Growth: Schumpeter Might Be Right." *Quarterly Journal of Economics* 108(3): 717–37.

References

Kitschelt, Herbert, Peter Lange, Gary Marks, and John D. Stephens, ed. 1999. *Continuity and Change in Contemporary Capitalism*. New York, NY: Cambridge University Press.

Klein, Benjamin. 2002. "The Role of Incomplete Contracts in Self-Enforcing Relationships." In *The Economics of Contracts: Theories and Applications*, ed. Eric Brousseau and Jean-Michel Glachant, 59–71. Cambridge: Cambridge University Press.

Klein, Benjamin, Robert Crawford, and Armen Alchian. 1996. "Vertical Integration, Appropriable Rents, and the Competitive Contracting Process." In *The Economic Nature of the Firm: A Reader*, ed. Louis Putterman and Randall S. Kroszner, 105–24. Cambridge: Cambridge University Press.

Kogut, Bruce, ed. 1993. *Country Competitiveness: Technology and the Organizing of Work*. New York, NY: Oxford University Press.

Kondratieff, Nikolai. 1984. *The Long Wave Cycle*. New York, NY: Richardson and Snyder.

Korkeamäki, Timo, and Tuomas Takalo. 2012. "Valuation of Innovation: The Case of iPhone." Bank of Finland Research Discussion Papers 24.

Kormendi, Roger, and Philip Meguire. 1985. "Macroeconomic Determinants of Economic Development: Cross-Country Evidence." *Journal of Monetary Economics* 16(2): 141–63.

Kornai, János. 1980. *Economics of Shortage*. Amsterdam: North-Holland.

Kornai, János. 1992. *The Socialist System: The Political Economy of Communism*. Princeton, NJ: Princeton University Press.

Korpi, Walter, and Joakim Palme. 1998. "The Paradox of Redistribution and Strategies of Inequality: Welfare State Institutions, Inequality, and Poverty in the Western Countries." *American Sociological Review* 63(5): 661–87.

Krause, George, and Kevin Corder. 2007. "Explaining Bureaucratic Optimism: Theory and Evidence from U.S. Executive Agency Macroeconomic Forecasts." *American Political Science Review* 101(1): 129–42.

Krauss, Gerhard, and Hans-Georg Wolf. 2002. "Technological Strengths in Mature Sectors: An Impediment or an Asset for Regional Economic Restructuring? The Case of Multimedia and Biotechnology in Baden-Württemberg." *Journal of Technology Transfer* 27(1): 39–50.

Kremer, Michael. 1993. "The O-ring Theory of Economic Development." *Quarterly Journal of Economics* 108(3): 551–75.

Kriesi, Hanspeter. 1982. "The Structure of the Swiss Political Systems." In *Patterns of Corporatist Policy-Making*, ed. Gerhard Lehmbruch and Phillippe Schmitter, 133–62. London: Sage.

Kristensen, Peer Hull, and Kari Lilja. 2011. *Nordic Capitalisms and Globalization: New Forms of Economic Organization and Welfare Institutions*. New York, NY: Oxford University Press.

Kuznets, Simon. 1955. "Economic Growth and Income Inequality." *American Economic Review* 45(1): 1–28.

Kuznets, Simon. 1963. "Quantitative Aspects of the Economic Growth of Nations, VIII: The Distribution of Income by Size." *Economic Development and Cultural Change* 11(2): 1–80.

La Porta, Rafael, Florencio Lopez-de-Silanes, Andrei Shleifer, and Robert Vishny. 1998. "Law and Finance." *Journal of Political Economy* 106(6): 1113–55.

References

Lafontaine, Francine, and Emmanuel Raynaud. 2002. "Residual Claims and Self-Enforcement as Incentive Mechanisms in Franchise Contracts: Substitutes or Complements?" In *The Economics of Contracts: Theories and Applications*, ed. Eric Brousseau and Jean-Michel Glachant, 315–36. Cambridge: Cambridge University Press.

Lamoreaux, Naomi, and Kenneth Sokoloff, ed. 2007. *Financing Innovation in the United States, 1870 to the Present.* Cambridge, MA: MIT Press.

Lancaster, Kelvin. 1982. "Innovative Entry: Profit Hidden Beneath the Zero." *Journal of Industrial Economics* 31(1–2): 41–56.

Landau, Ralph, Timothy Taylor, and Gavin Wright, ed. 1996. *The Mosaic of Economic Growth.* Stanford, CA: Stanford University Press.

Landes, David. 1969. *Unbound Prometheus: Technological Change and Industrial Development in Western Europe from 1750 to the Present.* New York, NY: Cambridge University Press.

Lane, Cristel, and Reinhard Bachmann. 1996. "The Social Constitution of Trust: Supplier Relations in Britain and Germany." *Organizational Studies* 17(3): 365–95.

Lange, Knut. 2009. "Institutional Embeddedness and the Strategic Leeway of Actors: The Case of the German Therapeutical Biotech Industry." *Socio-Economic Review* 7(2): 181–207.

Lau, Lawrence. 1996. "The Sources of Long-Term Economic Growth: Observations from the Experience of Developed and Developing Countries." In *The Mosaic of Economic Growth*, ed. Ralph Landau, Timothy Taylor, and Gavin Wright, 63–91. Stanford, CA: Stanford University Press.

Lawler, Edward, Susan Mohrman, and Gerald Ledford. 1995. *Creating High Performance Organizations: Practices and Results of Employee Involvement and Total Quality Management in Fortune 1000 Companies.* San Francisco, CA: Jossey-Bass.

Lawton-Smith, Helen. 2006. *Universities, Innovation, and the Economy.* New York, NY: Routledge.

Lazear, Edward, and Sherwin Rosen. 1981. "Rank-order Tournaments as Optimum Labor Contracts." *Journal of Political Economy* 89(5): 841–64.

Lazonick, William. 1990. *Competitive Advantage on the Shop Floor.* Cambridge, MA: Harvard University Press.

Lazonick, William. 2005. "The Innovative Firm." In *The Oxford Handbook of Innovation*, ed. Jan Fagerberg, David Mowery, and Richard Nelson, 29–55. New York, NY: Oxford University Press.

Lazonick, William. 2007. "Varieties of Capitalism and Innovative Enterprise." *Comparative Social Research* 24: 21–69.

Lazonick, William. 2010. "Innovative Business Models and Varieties of Capitalism: Financialization of the US Corporation." *Business History Review* 84(4): 675–702.

Leblang, David, and Bumba Mukherjee. 2004. "Presidential Elections and the Stock Market: Comparing Markov-Switching and Fractionally Integrated GARCH Models of Volatility." *Political Analysis* 12(3): 296–322.

Leblang, David, and Bumba Mukherjee. 2005. "Government Partisanships, Elections, and the Stock Market: Examining American and British Stock Returns, 1930–2000." *American Journal of Political Science* 49(4): 780–802.

Lebo, Matthew, and Janet Box-Steffensmeier. 2008. "Dynamic Conditional Correlations in Political Science." *American Journal of Political Science* 52(3): 688–704.

References

Lehmbruch, Gerhard. 1984. "Concertation and the Structure of Corporatist Networks." In *Order and Conflict in Contemporary Capitalism: Studies in the Political Economy of Western European Nations*, ed. John Goldthorpe, 60–80. Oxford: Oxford University Press.

Lerner, Josh, and Ulrike Malmendier. 2010. "Contractibility and the Design of Research Agreements." *American Economic Review* 100(1): 214–46.

Levi, Margaret. 1997. *Consent, Dissent, and Patriotism*. New York, NY: Cambridge University Press.

Levi, Margaret. 1988. *Of Rule and Revenue*. Berkeley, CA: University of California Press.

Levy, Jonah. 1999. *Tocqueville's Revenge: State, Society, and Economy in Contemporary France*. Cambridge, MA: Harvard University Press.

Levy, Jonah, ed. 2006. *The State after Statism: New State Activities in the Age of Liberalization*. Cambridge, MA: Harvard University Press.

Loriaux, Michael. 1991. *France after Hegemony: International Change and Financial Reform*. Ithaca, NY: Cornell University Press.

Luhmann, Niklas. 1968. *Vertrauen: Ein Mechanismus der Reduktion sozialer Komplexität*, 4th edn. Stuttgart: Lucius & Lucius, 2000.

Lundvall, Bengt-Åke, ed. 2010. *National Systems of Innovation: Toward a Theory of Innovation and Interactive Learning*. London: Anthem Press.

Lundvall, Bengt-Åke, and Edward Lorenz. 2011. "Social Investment in the Globalizing Learning Economy: A European Perspective." In *Towards a Social Investment Welfare State? Ideas, Policies and Challenges*, ed. Nathalie Morel, Bruno Palier, and Joakim Palme, 235–60. Bristol: The Policy Press.

Lundvall, Johannes. 2013. "Union Density and Political Strikes." *World Politics* 65(3): 539–69.

Lupu, Noam, and Jonas Pontusson. 2011. "The Structure of Inequality and the Politics of Redistribution." *American Political Science Review* 105(2): 316–66.

Lynch, Julia. 2006. *Age in the Welfare State: The Origins of Social Spending on Pensioners, Workers, and Children*. New York, NY: Cambridge University Press.

Lynch, Lisa, ed. 1994. *Training and the Private Sector: International Comparisons*. Chicago, IL: University of Chicago Press.

MacCulloch, Robert. 2004. "The Impact of Income on the Taste for Revolt." *American Journal of Political Science* 48(4): 830–48.

MacLeod, W. Bentley. 2002. "Complexity and Contract." In *The Economics of Contracts: Theories and Applications*, ed. Eric Brousseau and Jean-Michel Glachant, 213–41. Cambridge: Cambridge University Press.

Malerba, Franco. 1992. "Learning by Firms and Incremental Technical Change." *The Economic Journal* 102(413): 845–59.

Mansfield, Edwin. 1995. "Academic Research Underlying Industrial Innovations." *Review of Economics and Statistics* 77(1): 55–65.

Mares, Isabela. 2003. *The Politics of Social Risk: Business and Welfare State Development*. New York, NY: Cambridge University Press.

Mares, Isabela. 2004. "Wage Bargaining in the Presence of Social Services and Transfers." *World Politics* 57(1): 99–142.

References

Mares, Isabela. 2006. *Taxation, Wage Bargaining, and Unemployment*. New York, NY: Cambridge University Press.

Martin, Cathie Jo, and Jette Steen Knudsen. 2010. "Scenes from a Mall: Retail Training and the Social Exclusion of Low-Skilled Workers." *Regulation & Governance* 4(3): 345–64.

Martin, Cathie Jo, and Duane Swank. 2004. "Does the Organization of Capital Matter? Employers and Active Labor Market Policy at the National and Firm Levels." *American Political Science Review* 98(4): 593–611.

Martin, Cathie Jo, and Duane Swank. 2008. "The Political Origins of Coordinated Capitalism: Business Organizations, Party Systems, and the State Structure in the Age of Innocence." *American Political Science Review* 102(2): 181–98.

Martin, Cathie Jo, and Duane Swank. 2011. "Gonna Party Like It's 1899: Party Systems and the Origins of Varieties of Coordination." *World Politics* 63(1): 78–114.

Martin, Cathie Jo, and Duane Swank. 2012. *The Political Construction of Business Interests: Coordination, Growth, and Equality*. New York, NY: Cambridge University Press.

Martinelli, Alberto, ed. 1991. *International Markets and Global Firms: A Comparative Study of Organized Business in the Chemical Industry*. London: Sage.

Maxfield, Sylvia, and Ben Ross Schneider, ed. 1997. *Business and the State in Developing Countries*. Ithaca, NY: Cornell University Press.

Marx, Karl. 1867. *Das Kapital, Kritik der Politischen Ökonomie*. Chicago, IL: H. Regnery.

Mayer, Colin, and Xavier Vives, ed. 1993. *Capital Markets and Financial Intermediation*. New York, NY: Cambridge University Press.

McCall, Leslie. 2013. *The Undeserving Rich: American Beliefs about Inequality, Opportunity, and Redistribution*. New York, NY: Cambridge University Press.

Merges, Robert P., and Richard R. Nelson. 1994. "On Limiting or Encouraging Rivalry in Technical Progress: The Effect of Patent Scope Decisions." *Journal of Economic Behavior and Organization* 25(1): 1–24.

Milgrom, Paul, and Bengt Holmström. 1994. "The Firm as an Incentive System." *American Economic Review* 84(4): 972–91.

Miwa, Yoshiro, and Mark Ramseyer. 2006. *The Fable of the Keiretsu: Urban Legends of the Japanese Economy*. Chicago, IL: University of Chicago Press.

Moene, Karl Ove, and Michael Wallerstein. 2003. "Earnings Inequality and Welfare Spending: A Disaggregated Analysis." *World Politics* 55(4): 485–516.

Moffitt, Robert, and Peter Gottschalk. 2002. "Trends in the Transitory Variance of Earnings in the United States." *The Economic Journal* 112(478): C68–73.

Mokyr, Joel. 1990. *The Levers of Riches: Technological Creativity and Economic Progress*. Oxford: Oxford University Press.

Mokyr, Joel. 2002. *The Gifts of Athena: Historical Origins of the Knowledge Economy*. Princeton, NJ: Princeton University Press.

Moloney, Niamh. 2012. "The Legacy Effects of the Financial Crisis on Regulatory Design in the EU." In *The Regulatory Aftermath of the Global Financial Crisis*, ed. Eilís Ferran, Niamh Moloney, Jennifer G. Hill, and John C. Coffee, 111–202. New York, NY: Cambridge University Press.

Moran, Michael. 2006. "The Transformation of the British State: From Club Government to State-Administered High Modernism." In *The State after Statism: New State*

Activities in the Age of Liberalization, ed. Jonah Levy, 31–56. Cambridge, MA: Harvard University Press.

Morel, Nathalie, Bruno Palier, and Joakim Palme, ed. 2011. *Towards a Social Investment Welfare State? Ideas, Policies and Challenges*. Bristol: The Policy Press.

Müller, Walter, and Markus Gangl, ed. 2003. *Transitions from Education to Work in Europe: The Integration of Youth into EU Labor Markets*. Oxford: Oxford University Press.

Müller, Walter, and Wolfgang Karle. 1993. "Social Selection and Educational System in Europe." *European Sociological Review* 9(1): 1–23.

Müller, Walter, and Yossi Shavit. 1998. "The Institutional Embeddedness of the Stratification Process: A Comparative Study of Qualifications and Occupations in Thirteen Countries." In *From School to Work: A Comparative Study of Educational Qualifications and Occupational Destinations*, ed. Yossi Shavit and Walter Müller, 1–48. Oxford: Clarendon Press.

Nelson, Moira, and John D. Stephens. 2012. "Do Social Investment Policies Produce More and Better Jobs?" In *Towards a Social Investment Welfare State? Ideas, Policies and Challenges*, ed. Nathalie Morel, Bruno Palier, and Joakim Palme, 205–34. Bristol: The Policy Press.

Nelson, Phillip. 1970. "Information and Consumer Behavior." *Journal of Political Economy* 78(2): 311–29.

Nelson, Richard. 1993. *National Innovation Systems: A Comparative Analysis*. Cambridge, MA: Harvard University Press.

Nelson, Richard. 1996. *The Sources of Economic Growth*. Cambridge, MA: Harvard University Press.

Nelson, Richard. 2005. *Technology, Institutions, and Economic Growth*. Cambridge, MA: Harvard University Press.

Nielsen, Richard, Michael Findley, Zachary Davis, Tara Candland, and Daniel Nielson. 2011. "Foreign Aid Shocks as a Cause of Violent Armed Conflict." *American Journal of Political Science* 55(2): 219–32.

Noble, Gregory. 1998. *Collective Action in East Asia: How Ruling Parties Shape Industrial Policy*. Ithaca, NY: Cornell University Press.

Nye, David. 2013. *America's Assembly Line*. Cambridge, MA: MIT Press.

O'Sullivan, Mary. 2000. *Contests for Corporate Control: Corporate Governance and Economic Performance in the United States and Germany*. New York, NY: Oxford University Press.

Okun, Arthur. 1975. *Equality and Efficiency: The Big Tradeoff*. Washington, DC: The Brookings Institute.

Olson, Mancur. 1965. *The Logic of Collective Action: Public Goods and the Theory of Groups*. Cambridge, MA: Harvard University Press.

Olson, Mancur. 2000. *Power and Prosperity: Outgrowing Communist and Capitalist Dictatorships*. New York, NY: Basic Books.

Organization for Economic Cooperation and Development. 1994. *The OECD Jobs Study*. Paris: OECD.

Organization for Economic Cooperation and Development. 1995. *National Systems for Financing Innovation*. Paris: OECD.

Organization for Economic Cooperation and Development. 1996a. *Technology, Productivity, and Job Creation: Analytical Report*. Paris: OECD.

References

Organization for Economic Cooperation and Development. 1996b. *Technology and Industrial Performance: Technology Diffusion, Productivity, Employment and Skills, International Competitiveness*. Paris: OECD.

Organization for Economic Cooperation and Development. 1999. *OECD Employment Outlook*. Paris: OECD.

Organization for Economic Cooperation and Development. 2001. *Measuring Productivity: Measurement of Aggregate and Industry-Level Productivity Growth*. Paris: OECD.

Organization for Economic Cooperation and Development. 2002a. *OECD Employment Outlook*. Paris: OECD.

Organization for Economic Cooperation and Development. 2002b. *Frascati Manual: Proposed Standard Practice for Surveys on Research and Experimental Development*. Paris: OECD.

Organization for Economic Cooperation and Development. 2005. *Oslo Manual: Guidelines for Collecting and Interpreting Innovation Data*. Paris: OECD.

Organization for Economic Cooperation and Development. 2010. *OECD Science, Technology and Industry Outlook*. Paris: OECD.

Ornston, Darius. 2012. *When Small States Make Big Leaps: Institutional Innovation and High-Tech Competition in Western Europe*. Ithaca, NY: Cornell University Press.

Ornston, Darius. 2013. "Creative Corporatism: The Politics of High-Technology Competition in Nordic Europe." *Comparative Political Studies* 46(6): 702–29.

Oster, Emily, Ira Shoulson, and E. Ray Dorsey. 2013. "Limited Life Expectancy, Human Capital and Health Investments." *American Economic Review* 103(5): 1977–2002.

Pagano, Marco, and Paolo Volpin. 2005. "The Political Economy of Corporate Governance." *American Economic Review* 95(4): 1005–30.

Palier, Bruno, and Kathleen Thelen. 2010. "Institutionalizing Dualism: Complementarities and Change in France and Germany." *Politics & Society* 38(1): 119–48.

Penrose, Edith. 1995. *The Theory of the Growth of the Firm* 3rd edn. Oxford: Oxford University Press.

Phlips, Louis, and Jacques-François Thisse. 1982. "Spatial Competition and the Theory of Differentiated Markets: An Introduction." *Journal of Industrial Economics* 31(1–2): 1–9.

Pianta, Mario. 1998. "New Technology and Jobs." In *Globalization, Growth, and Governance: Creating an Innovative Economy*, ed. Jonathan Michie and John Grieve Smith, 71–100. New York, NY: Oxford University Press.

Pianta, Mario. 2005. "Innovation and Employment." In *The Oxford Handbook of Innovation*, ed. Jan Fagerberg, David Mowery, and Richard Nelson, 568–98. New York, NY: Oxford University Press.

Pigou, Arthur. 1912. *Wealth and Welfare*. London: Macmillan.

Piore, Michael, and Charles Sabel. 1984. *The Second Industrial Divide: Possibilities for Prosperity*. New York, NY: Basic Books.

Pissarides, Christopher. 1997. "Learning by Trading and the Returns to Human Capital in Developing Countries." *World Bank Economic Review* 11: 17–32.

Pitelis, Christos, ed. 2002. *The Growth of the Firm: The Legacy of Edith Penrose*. New York, NY: Oxford University Press.

References

Piva, Mariacristina, Enrico Santarelli, and Marco Vivarelli. 2005. "The Skill Bias Effect of Technological and Organizational Change: Evidence and Policy Implications." *Research Policy* 34(2): 141–57.

Pontusson, Jonas. 2005. *Inequality and Prosperity: Social Europe vs. Liberal America*. Ithaca, NY: Cornell University Press.

Powell, Walter, and Stine Grodal. 2005. "Networks of Innovators." In *The Oxford Handbook of Innovation*, ed. Jan Fagerberg, David Mowery, and Richard Nelson, 56–85. New York, NY: Oxford University Press.

Pryor, Frederic. 1968. *Public Expenditures in Communist and Capitalist Nations*. Homewood, IL: Unwin.

Rafiqui, Pernilla. 2010. "Varieties of Capitalism and Local Outcomes: A Swedish Case Study." *European Urban and Regional Studies* 17(3): 309–29.

Raith, Michael. 2003. "Competition, Risk, and Managerial Incentives." *American Economic Review* 93(4): 1425–36.

Rajan, Raghuram, and Luigi Zingales. 2001. "The Firm as a Dedicated Hierarchy: A Theory of the Origins and Growth of Firms." *Quarterly Journal of Economics* 116(3): 805–51.

Regini, Marino, and Gøsta Esping-Andersen, ed. 2000. *Why Deregulate Labor Markets?* New York, NY: Oxford University Press.

Rehm, Philipp. 2011. "Social Policy by Popular Demand." *World Politics* 63(2): 271–99.

Rehm, Philipp, Jacob S. Hacker, and Mark Schlesinger. 2012. "Insecure Alliances: Risk, Inequality, and Support for the Welfare State." *American Political Science Review* 106(2): 386–406.

Reitzig, Markus. 2004. "The Private Values of 'Thickets' and 'Fences': Towards an Updated Picture of the Use of Patents across Industries." *Economics of Innovation and New Technology* 13(5): 457–76.

Roberts, Kenneth, and Erik Wibbels. 1999. "Party Systems and Electoral Volatility in America: A Test of Economic, Institutional, and Structural Explanations." *American Political Science Review* 93(3): 575–90.

Robertson, Graeme. 2007. "Strikes and Labor Organization in Hybrid Regimes." *American Political Science Review* 101(4): 781–98.

Roe, Mark. 2002. *Political Determinants of Corporate Governance*. Oxford: Oxford University Press.

Rolfe, Meredith. 2012. *Voter Turnout: A Social Theory of Political Participation*. New York, NY: Cambridge University Press.

Rosen, Sherwin. 1976. "A Theory of Life Earnings." *Journal of Political Economy* 84(4): S45–67.

Rosenberg, Nathan. 1996. "Uncertainty and Technological Change." In *The Mosaic of Economic Growth*, ed. Ralph Landau, Timothy Taylor, and Gavin Wright, 334–57. Stanford, CA: Stanford University Press.

Rueda, David. 2005. "Insider–Outsider Politics in Industrialized Democracies: The Challenge to Social Democratic Parties." *American Political Science Review* 99(1): 61–74.

Rueda, David. 2007. *Social Democracy Inside Out: Partisanship and Labor Market Policy in Industrialized Democracies*. New York, NY: Oxford University Press.

References

Rueda, David. 2008. "Left Government, Policy, and Corporatism: Explaining the Influence of Partisanship on Inequality." *World Politics* 60(3): 349–89.

Rumberger, Russell. 1981. *Overeducation in the US Labor Market*. New York, NY: Praeger.

Sabel, Charles, and Jonathan Zeitlin, ed. 1997. *World of Possibilities: Flexible and Mass Production in Western Industrialization*. New York, NY: Cambridge University Press.

Sainsbury, Diane, ed. 1999. *Gender and Welfare State Regimes*. New York, NY: Oxford University Press.

Saka, Ayse. 2002. "Institutional Limits to the Internalization of Work Systems: A Comparative Study of Three Japanese Multinational Companies in the UK." *European Journal of Industrial Relations* 8(3): 251–75.

Sako, Mari. 1992. *Prices, Quality and Trust: Inter-firm Relations in Britain and Japan*. New York, NY: Cambridge University Press.

Saxenian, Annalee. 1994. *Regional Advantage: Culture and Competition in Silicon Valley and Route 128*. Cambridge, MA: Harvard University Press.

Scharpf, Fritz, and Vivien Schmidt, ed. 2000. *Welfare and Work in the Open Economy Volume II: Diverse Response to Common Challenges in Twelve Countries*. New York, NY: Oxford University Press.

Scheve, Kenneth, and Matthew Slaughter. 2004. "Economic Insecurity and the Globalization of Production." *American Journal of Political Science* 48(4): 662–74.

Schmidt, Vivien. 1990. *Democratizing France: The Political and Administrative History of Decentralization*. New York, NY: Cambridge University Press.

Schmitz, Patrick. 2006. "Information Gathering, Transaction Costs, and the Property Rights Approach." *American Economic Review* 96(1): 422–34.

Schneiberg, Marc. 2007. "What's on the Path? Path Dependence, Organizational Diversity, and the Problem of Institutional Change in the US Economy, 1900–1950." *Socio-Economic Review* 5(1): 47–80.

Schumpeter, Joseph. 1942. *Capitalism, Socialism, and Democracy*. London: Harper and Brothers.

Shaked, Avner, and John Sutton. 1987. "Product Differentiation and Industrial Structure." *Journal of Industrial Economics* 36(2): 131–46.

Shapiro, Carl, and Joseph Stiglitz. 1984. "Equilibrium Unemployment as a Worker Discipline Device." *American Economic Review* 74(3): 433–44.

Shavit, Yossi, and Walter Müller, ed. 1998. *From School to Work: A Comparative Study of Educational Qualifications and Occupational Destinations*. Oxford: Clarendon Press.

Siebert, Horst. 1997. "Labor Market Rigidities: At the Root of Unemployment in Europe." *Journal of Economic Perspectives* 11(3): 37–54.

Simon, Herbert. 1951. "A Formal Theory of the Employment Relationship." *Econometrica* 19(3): 293–305.

Smitka, Michael. 1991. *Competitive Ties: Subcontracting in the Japanese Automotive Industry*. New York, NY: Columbia University Press.

Solow, Robert. 1957. "Technical Change and the Aggregate Production Function." *Review of Economics and Statistics* 39(3): 312–20.

Solt, Frederick. 2008. "Economic Inequality and Democratic Political Engagement." *American Journal of Political Science* 52(1): 48–60.

References

Soskice, David. 1994. "Reconciling Markets and Institutions: The German Apprenticeship System." In *Training and the Private Sector: International Comparisons*, ed. Lisa Lynch, 25–60. Chicago, IL: University of Chicago Press.

Spence, Michael. 1973. "Job Market Signaling." *Quarterly Journal of Economics* 87(3): 355–74.

Spiezia, Vincenzo, and Marco Vivarelli. 2002. "Innovation and Employment: A Critical Survey." In *Productivity, Inequality, and the Digital Economy: A Transatlantic Perspective*, ed. Nathalie Greenan, Yannick L'Horty, and Jacques Mairesse, 101–32. Cambridge, MA: MIT Press.

Stasavage, David. 2005. "Democracy and Education Spending in Africa." *American Journal of Political Science* 49(2): 343–58.

Steil, Benn, David Victor, and Richard Nelson, ed. 2002. *Technological Innovation and Economic Performance*. Princeton, NJ: Princeton University Press.

Steinmo, Sven. 2010. *The Evolution of Modern States: Sweden, Japan, and the United States*. New York, NY: Cambridge University Press.

Stephens, John D. 2012. "Social Investment and the Knowledge Economy." *Socio-Economic Review* 10(4): 765–71.

Stevens, Margaret. 1996. "Transferrable Training and Skill Poaching." In *Acquiring Skills: Market Failures, Their Symptoms and Policy Responses*, ed. Alison Booth and Dennis Snower, 21–40. London: Centre for Economic Policy Research.

Stiglitz, Joseph, Daniel McFadden, and Sam Peltzman. 1987. "Technological Change, Sunk Costs, and Competition." *Brookings Papers on Economic Activity* 3: 883–947.

Stiglitz, Joseph, and Andrew Weiss. 1981. "Credit Rationing in Markets with Imperfect Information." *American Economic Review* 71(3): 393–410.

Streeck, Wolfgang. 1989. "Skills and the Limits of Neo-liberalism." *Work, Employment & Society* 3(1): 90–104.

Streeck, Wolfgang, ed. 2005. *Governing Interests: Business Associations Facing Internationalization*. New York, NY: Routledge.

Streeck, Wolfgang, and Kathleen Thelen, ed. 2005. *Beyond Continuity: Institutional Change in Advanced Political Economies*. New York, NY: Oxford University Press.

Sutton, John. 1986. "Vertical Product Differentiation: Some Basic Themes." *American Economic Review* 76(2): 393–8.

Swank, Duane. 2002. *Global Capital, Political Institutions, and Policy Change in Developed Welfare States*. New York, NY: Cambridge University Press.

Tafara, Ethiopis. 2012. "Foreword." In *The Regulatory Aftermath of the Global Financial Crisis*, ed. Eilís Ferran, Niamh Moloney, Jennifer G. Hill, and John C. Coffee, xi–xxvi. New York, NY: Cambridge University Press.

Tandelis, Steven. 2002. "Complexity, Flexibility, and the Make-or-Buy Decision." *American Economic Review* 92(2): 433–7.

Taylor, Mark Zachary. 2004. "Empirical Evidence against Varieties of Capitalism's Theory of Technological Innovation." *International Organization* 58(3): 601–30.

Teece, David. 1980. "Economies of Scope and the Scope of the Enterprise." *Journal of Economic Behavior and Organization* 1(3): 223–47.

Thelen, Kathleen. 2004. *How Institutions Evolve: The Political Economy of Skills in Germany, Britain, the United States, and Japan*. New York, NY: Cambridge University Press.

References

Thelen, Kathleen. 2014. *Varieties of Liberalization: The New Politics of Social Solidarity*. New York, NY: Cambridge University Press.

Thomas, Damian. 2011. "Ireland: Two Trajectories of Institutionalization." In *Social Pacts in Europe: Emergence, Evolution, and Institutionalization*, ed. Sabina Avdagic, Martin Rhodes, and Jelle Visser, 89–117. New York, NY: Oxford University Press.

Thomson Reuters. 2012. *National Venture Capital Association Yearbook 2012*. New York, NY.

Thurow, Lester. 1976. *Generating Inequality: Mechanisms of Distribution in the US Economy*. London: Macmillan Press.

Tirole, Jean. 2009. "*Cognition and Incomplete Contracts*." American Economic Review 99(1): 265–94.

Tödtling, Franz, and Alexander Kaufmann. 2002. "SMEs in Regional Innovation Systems and the Role of Innovation Support: The Case of Upper Austria." *Journal of Technology Transfer* 27(1): 15–26.

Trigilia, Carlo, and Luigi Burroni. 2009. "Italy: Rise, Decline and Restructuring of a Regionalized Capitalism." *Economy and Society* 38(4): 630–53.

Trumbull, Gunnar. 2004. *Silicon and the State: French Innovation Policy in the Internet Age*. Washington, DC: Brookings Institute Press.

Tuck, Richard. 2008. *Free Riding*. Cambridge, MA: Harvard University Press.

Tylecote, Andrew. 1996. "Managerial Objectives and Technological Collaboration: The Role of National Variations in Cultures and Structures." In *Technological Collaboration: The Dynamics of Cooperation in Industrial Innovation*, ed. Rod Commbs, 34–53. Cheltenham, UK: Edward Elgar.

United States Census Bureau. 2012. Educational Attainment in the United States: 2012 Detailed Tables (<http://www.census.gov/hhes/socdemo/education/data/cps/2012/tables.html>).

Utterback, James. 1994. *Mastering the Dynamics of Innovation: How Companies Can Seize Opportunities in the Face of Technological Change*. Cambridge, MA: Harvard Business School Press.

Vail, Mark. 2009. "Bending the Rules: Institutional Analysis, Political Change and Labor Market Reform in Advanced Industrial Societies." *Comparative Politics* 42(1): 21–39.

Van den Steen, Eric. 2010. "Interpersonal Authority in a Theory of the Firm." *American Economic Review* 100(1): 466–90.

van Reenen, John. 1996. "The Creation and Capture of Rents: Wages and Innovation in a Panel of U.K. Companies." *Quarterly Journal of Economics* 111(1): 195–226.

Verdier, Daniel. 2002. *Moving Money: Banking and Finance in the Industrialized World*. New York, NY: Cambridge University Press.

Verdugo, Richard, and Naomi B. Verdugo. 1989. "The Impact of Surplus Schooling on Earnings: Some Additional Findings." *Journal of Human Resources* 24(4): 629–43.

Visser, Jelle, and Anton Hemerijck. 1997. *A Dutch Miracle: Job Growth, Welfare Reform and Corporatism in the Netherlands*. Amsterdam: University of Amsterdam Press.

Vivarelli, Marco, Ronaldo Evangelista, and Mario Pianta. 1996. "Innovation and Employment in Italian Manufacturing Industry." *Research Policy* 25(7): 1013–26.

Vivarelli, Marco, and Mario Pianta, ed. 2000. *The Employment Impact of Innovation: Evidence and Policy*. London: Routledge.

References

Waldfogel, Jane. 2002. "Child Care, Women's Employment, and Child Outcomes." *Journal of Population Economics* 15(3): 527–48.

Wallerstein, Michael. 2008. *Selected Works of Michael Wallerstein: The Political Economy of Inequality, Unions, and Social Democracy*, ed. David Austen-Smith. New York, NY: Cambridge University Press.

Watts, Rob. 1997. "Ten Years On: Frances G. Castles and the Australian 'Wage-Earners' Welfare State." *Journal of Sociology* 33(1): 1–15.

Weitzman, Martin. 1998. "Recombinant Growth." *Quarterly Journal of Economics* 113(2): 331–60.

Western, Bruce. 1997. *Between Class and Market: Postwar Unionization in the Capitalist Democracies*. Princeton, NJ: Princeton University Press.

Whitley, Richard. 1999. *Divergent Capitalisms: The Social Structuring and Change of Business Systems*. Oxford: Oxford University Press.

Whittaker, D. Hugh. 1997. *Small Firms in the Japanese Economy*. New York, NY: Cambridge University Press.

Widmer, Frédéric. 2011. "Institutional Investors, Corporate Elites and the Building of a Market for Corporate Control." *Socio-Economic Review* 9(4): 671–97.

Wilensky, Harold, and C. N. Lebeaux. 1958. *Industrial Society and Social Welfare*. New York, NY: Sage.

Williamson, Oliver. 1985. *The Economic Institutions of Capitalism: Firms, Markets, Relational Contracting*. New York, NY: Free Press.

Wilson, Harold. October 1, 1963. "Labour's Plan for Science." Speech by Harold Wilson at the Annual Conference of the British Labour Party, Scarborough, UK.

Wolff, Edward. 2000. "Technology and the Demand for Skills." In *The Overeducated Worker? The Economics of Skill Utilization*, ed. Lex Borghans and Andries de Grip, 27–57. Northampton, MA: Edward Elgar.

Wren, Anne, ed. 2012. *The Political Economy of the Service Transition*. New York, NY: Oxford University Press.

Zeira, Joseph. 1998. "Workers, Machines, and Economic Growth." *Quarterly Journal of Economics* 113(4): 1091–117.

Zeitlin, Jonathan, and David Trubek, ed. 2003. *Governing Work and Welfare in a New Economy: European and American Experiments*. New York, NY: Oxford University Press.

Zellner, Arnold. 1962. "An Efficient Method of Estimating Seemingly Unrelated Regression Equations and Tests for Aggregation Bias." *Journal of the American Statistical Association* 57 (298): 348–68.

Zysman, John. 1994. "How Institutions Create Historically Rooted Trajectories of Growth." *Industrial and Corporate Change* 3(1): 243–83.

Index

Academic education 108, 110–12, 117–21, 227
Adverse selection 14, 18–20, 33–40
Allocative efficiency 105
Assembly line 119
Asset specific human capital 114–17
Austria
  small and median enterprises 72–3
  subcontracting 186–7

Banking 3, 37, 50–3, 220–2
Buyout 46, 221

Collective action failure 23, 81
Communication 11, 69, 115
Community Innovation Survey 65–6, 180
Control and discretion 7–8, 167–9, 178, 183–8, 211–12
Creative corporatism 31, 193, 215–16
Credit rationing 3, 12, 14, 17–21, 50–1

Demotivation 8–9, 170, 189–226
Dualism 60–1, 88, 92–4, 98, 108, 233
Dynamic efficiency 105–6, 136, 232

Economic volatility 7–8, 164–7, 193–6, 200–10
Education systems 125–6, 132, 228–9
Electoral systems 115–16
Embodied and disembodied knowledge 67–8, 198–9
Employee Stock Ownership Plan 226
Employment 81

Financial market regulation 8, 218–23

Germany
  banks 55–6
  biotechnology 55, 73–4, 138, 193
  media industry 55, 75
  universities 129
Globalization
  capital liberalization 147
  global financial crisis 46–8, 218–19
  international competition 58, 61, 73
  trade openness 33, 166–7
Growth accounting 96

High-powered incentives 20, 170, 189–91, 226

Inequality
  educational inequality 6, 107–12, 121, 132–5
  income inequality 6, 9, 107–9, 113, 141–2, 147–8, 154–6
Initial Public Offerings 12, 18–19, 28–9, 33–9, 221–2
Innovation
  effect on employment 5, 59, 76–8, 82–95
  effect on inequality 107–9, 141–60, 230
  effect on productivity 5, 9, 61, 95, 99–104
  financing of innovation 10, 17–18
  incremental innovation 1, 6, 30, 62–3, 90–1, 105–6, 136–46, 193
  mutual licensing 182
  organizational innovation 69, 199
  process and product innovation 5, 59–67, 79, 104–5, 110, 197
  radical innovation 1, 6, 20, 30–1, 38, 62–3, 90–2, 105–10, 136–46, 193, 214–15
  regional innovation 75–6, 131–2
  research and development 92
  role of political parties 230–2
  role of the state 32, 68, 141, 199
  role of unions 33, 231–2
  science policy 231
  subcontracting 68, 167, 180–2, 186
  systems of innovation 69, 231
Insider and public information 3, 12–18, 55–7, 219–23
International Social Survey Program 122–3, 126
International Standard Classification of Education 124, 137
International Standard Classification of Occupations 122–3, 137
Italy
  biotechnology 74
  subcontracting 187
  Third Italy 22, 74, 187–8

Japan
  corporate governance 56–7
  workplace management 69, 186
Job market competition 107–8, 118–20, 162–3, 227

Index

Knowledge-intensive employment 214, 216–18, 225
Kuznets hypothesis 141–2, 145–6

Labor market policies 98, 217–18

Minority shareholder protection 28, 31–2
Monopolistic competition 77–8, 223, 228
Monopoly 64–5

Neocorporatism 22, 81, 166
Network externality 64–5
New social risks 164

Overeducation 6, 108–14, 117–27, 132–41, 161–2, 228

Patents 29, 85
Patient capital 23, 220
Predatory pricing 224
Principal–agent relationship 7, 167–8, 171–3
Productivity 61, 95–7, 117–20, 223–4
Property rights 168–9, 174–6
Public sector 98

Rank-order tournament 228
Rat race 3, 12, 17–21, 58, 118–19, 162–3, 227
Redistribution 230–1
Relational contracting 169, 173, 188
Residual ownership and control 169, 175–9, 190
Risk smoothing and pooling 15–17, 168

School-to-work transition 125, 228
Security Exchange Commission 219–20
Seemingly Unrelated Regression 69
Service industry 60–1, 88, 124, 144, 155, 218
Signaling 12–17, 31–2, 111, 116–20, 125
Skill training 108, 113–17, 136–7, 223, 229
Small Business Innovation Research Program 53–4

Social investment state 8, 214–15
State intervention 24, 98, 165
Stock markets 3, 12–13
Sweden
  banking 56
  furniture industry 56, 75–6
  software industry 74–5
Switzerland
  banking 56
  machine industry 56

Taxation 81
Theil's T 160
Total factor productivity 95–7
Trust 8, 170, 189, 191–2, 225–6
The United Kingdom
  biotechnology 55, 138
  software industry 74–5
  universities 129
  workplace management 186
The United States
  biotechnology 54, 73, 193
  financialization 154–5
  regional economy 188–9, 226–7

Varieties of capitalism 11, 22–4, 115, 174
  measurement 24–8
  Nordic variant 37–8, 135, 214–16
  ownership hierarchy 169, 174, 178
  role of the state 24, 169, 178, 213–18
  Southern European variant 26, 31, 37, 135
  subnational variation 57, 75, 131, 187–9
Venture capital 3, 12–13, 20, 29, 39–50, 53–5, 220–2
Vocational education 125

Wage bargaining 11, 80–1, 143–4, 175–6
Welfare states 9, 26, 215, 232–3
Workplace autonomy 224–5